Synergy

Peter Cornish • Gillian Berry
Editors

Synergy

The Healing Energy Guiding Stepped Care 2.0

Springer

Editors
Peter Cornish
Student Mental Health
University of California, Berkeley
Berkeley, CA, USA

Gillian Berry
Stepped Care Solutions
Mount Pearl, Canada

ISBN 978-3-031-88492-4 ISBN 978-3-031-88493-1 (eBook)
https://doi.org/10.1007/978-3-031-88493-1

This Springer imprint is published by the registered company Springer Nature Switzerland AG
The registered company address is: Gewerbestrasse 11, 6330 Cham, Switzerland

Acknowledgments

Land Acknowledgment (Nares Inlet)

Peter Cornish I honor the rocks, clear waters, trees, and islands of Eastern Georgian Bay that mean so much to me, my family, my ancestors, and, more importantly, the Indigenous peoples, who, for thousands of years, have been stewards of the land. With respect to this synergy project, I am grateful to Kay Scale, daughter-in-law to Marge and Nick Scale, who opened up a road to Nares Inlet 75 years ago. Kay is a member of the Shawanaga First Nation and introduced me to her best friend, Marylin Capriole, an elder of the Shawanaga First Nation, who offered a prayer at our talking circle. Marylin reminds all of us that the land doesn't belong to us. We are visitors who share it with all creatures and all generations, past and future.

Acknowledgment of Ancestors, Mentors, Colleagues, and Companions

Gillian Berry I will forever be grateful for the words, actions and, most importantly, the love of my ancestors who have transcended this earthly plane. I am who I am because of them, and I cannot deny their influence on my life. I am so grateful for the information they left me as a beacon of light to follow, and it is because of their example that I have been able to find my life's purpose.

Peter: Dick Katz is a kind and gentle man who, despite treading lightly in this world, remains focused on his commitment to sharing with the world the wisdom of ancient stories. Dick set me on a career path beyond psychology, one that is richer, deeper, more spiritual, and more community-based than I ever could have imagined. Because of Dick's influence, I see a whole other world of experience to tap into. And Dick, keeping his promise to share the stories of First Peoples, the true "First Psychologists," has opened strands of wisdom that I have long searched for.

Gillian: Thank you, Dick—for being curious and following wherever the breadcrumbs led you. Thank you for ignoring the voices that said otherwise and being dedicated to translating "unfamiliar concepts" into words I could understand. Tara, I appreciate your wise counsel, as it bridges the gap between two worlds. Thank you for seeing value in both and providing guidance for those who are to follow.

Peter: I am grateful for the trust granted me by my newest colleagues at UC Berkeley. Without knowing where they were going or really quite why, they traveled across the continent for a three-day retreat to experience synergy in a place that is near and dear to my heart. Thank you, Tobirus Newby, Matt Goodman, Adisa Anderson, Chris MacLean, Krist Panik, and David Presti, for joining us on this adventure.

Gillian: Integral to this book is the SCS documentary on Synergy and the work of Dick Katz. We must thank Mike and Sarah and all those at InsidersContent who willingly went above and beyond, both personally and professionally, to join us on our journey of discovery. As a result of recognizing the magnitude of the task—translating the concept of synergy onto film—we believe they have successfully captured the warmth, joy, and spiritual nature that is synergy.

Peter: AnnMarie Churchill, you are the steadfast partner who first joined me at a fork in the road 15 years ago and now leads us all on the ever-evolving SC2.0 journey of system transformation. Early in our journey, we recognized that what we were doing wasn't new. It has been and still is firmly rooted in ancient ways of practicing wellbeing.

Gillian: Striving to build a better world can sound so clichéd. Stepped Care Solutions, with its vision of wellbeing everywhere and its mission of changing the narrative in mental health systems, is truly about facing this seemingly impossible task. I am so thankful to be part of a curious, reflective team that faces challenging conundrums head-on. To have practical, emotional, and authentic support for this creative thought leadership project has been one of the most rewarding experiences of my life, for which my colleagues and I are truly grateful.

Peter: To our contributors, Raghav, Genée, Kareem, Catie, and Sarah—you brought disparate voices to these pages. Without your unique voices and identities, there would be no conundrums to power the healing energy of synergy.

Gillian: I am so grateful to those who worked with me at The George Washington University. Although there are too many to name here, I invite readers to listen to them directly as they share their experiences of implementing SC2.0 on our YouTube channel (https://www.youtube.com/@steppedcaresolutions). I know you will agree that our work in transforming the counseling center for the betterment of the students we served has also had a profound impact on our lives—the work we did lives on wherever we find ourselves. I will always be indebted to you all from the bottom of my heart for trusting the process.

Gillian: We do not think it is a coincidence, André, that your love of the Canadian wilderness, from which you learned so much, has found its way into this book. There is no doubt that your experiences in the wilderness have been beautifully interwoven into the words of this book. Thank you for your voice of reason, patience, and insights that kept us on track and gave life to our words.

Peter: Animals connect us with something larger than ourselves and to each other. Tinker, Spooky, Cheeky, Earnie, Henny Penny, Racky Racoon, Sparky, Frisky, Catnap, Cleo, Venice—these are creatures who blessed me with their gifts throughout their short lives. Your love brought family members together. We bonded over your antics. You connect us to something beyond ourselves. You also brought us our first experiences with grief and loss. Through you, we witnessed the power of returning to the land while remaining in our hearts.

Gillian: Thank you, Coco. Your personality shone through. I had to learn to be patient as you sniffed new smells. I valued your barking as you protected our home with a fierceness that made bigger dogs afraid despite the fact that you were only 3.5 lbs. I had to reluctantly accept your agency to choose what you liked to eat. Your presence made our lives better.

Peter & Gillian We extend our heartfelt gratitude to the cherished people in our lives—Karen, Nathaniel, Jenica, Heather, Harvey, Barb, Shakira, Patsy, Almina, and all those who know their role in this journey. Your patience, understanding, and unwavering support have meant the world to us. We honor and respect this great blue-green planet shrouded in strands of white, the forces and spirits that brought us into being, and the resilient ecosystems that gently nurture us with wonder—despite our missteps. To the lakes, rivers, oceans, creatures, rocks, plants, fields, mountains, and trees, we are of you, now and forever. We hope that our small contribution to the majesty of this world complements the goal of bringing wellbeing everywhere.

Contents

About the Editors

Gillian Berry is vice president of culture, diversity, and workforce development at Stepped Care Solutions. She is a published author and a licensed clinical social worker in the District of Columbia and the State of Maryland, with over 35 years of experience as an educator and practitioner. She has overseen the counseling and psychological services at George Washington University, leading a multidisciplinary team of mental health professionals through the implementation of the Stepped Care 2.0 approach into their service delivery. She has been a faculty member at the University of Botswana's Department of Social Work Southern Africa and the University of Maryland School of Social Work. Her clinical interests are strongly influenced by her grandfather's Indigenous philosophy, and she provides short-term solution-focused services/therapy from this perspective.

Peter Cornish Since developing the Stepped Care 2.0© (SC2.0) methodology, Peter Cornish has provided mental health system consultation and on-site implementation training to over 150 organizations across North and South America. In addition to his role as faculty and senior adviser at Stepped Care Solutions, Cornish is an honorary research professor at Memorial University and co-director of Student Mental Health (Strategic) at the University of California, Berkeley, where he established the Berkeley Center for Mental Health Innovation. He is collaborating with colleagues on a three-book series expanding on the SC2.0 methodology. He is a principal investigator for a $1.14 million Canadian Institutes of Health Research transitions-in-care, four-year research grant aimed at digitizing and evaluating SC2.0 across three Canadian provinces.

André Forget is a novelist, journalist, and literary critic. He is the author of *In the City of Pigs*, and editor of *After Realism: Twenty-four Stories for the Twenty-first Century.* His writing has appeared in many magazines and newspapers in Canada and the United States.

Catie Greene, PhD, LPC is a counselor educator and supervisor. Catie has worked as a clinician in college counseling centers and private practice, and an educator in

Clinical Mental Health Counseling graduate programs for over 11 years, and joined Stepped Care Solutions as a faculty consultant on synergy beyond diversity, equity, and inclusion in 2020. Catie's research and practice is grounded in Relational Cultural Theory. Catie recently launched Healing Stories, Animal Party Kiki Books, LLC, a small indie children's publishing company focused on creating therapeutic stories for children.

Sarah Harte, LCSW is a clinical social worker with over 20 years of experience working with young adults and their families. She has developed clinical and leadership skills in a variety of settings, including leadership roles in various universities and clinics. She knows the importance of working to build teams and a workplace culture where mental health professionals can enjoy satisfying, sustainable, and healthy careers. Sarah currently serves as a Program Development Manager with Stepped Care Solutions, working to transform mental health service delivery systems to foster wellbeing everywhere. Sarah holds a master's degree in Social Work and a post-master's certificate in Clinical Supervision from New York University. Sarah is an adjunct lecturer at Columbia University in New York and a part-time ceramic artist.

List of Figures

Chapter 1
What Is Synergy?

Peter Cornish

Introduction

I first encountered the concept of synergy as a graduate student at the University of Saskatchewan in the 1980s, when I enrolled in a course taught by Professor Richard Katz. Katz—who, then as now, insisted on going by the more informal "Dick,"—was an odd figure on campus. Though he had received his PhD in Clinical Psychology from Harvard, and went on to teach there for 20 years, he had come to feel that the Ivy League world was too focused on the tradition of scientific practice that had emerged in Europe during the seventeenth and eighteenth centuries. Having spent years living and working as an ethnographer in Namibia, Tahiti, Alaska, North Dakota, and Saskatchewan, he had come to realize that Indigenous philosophical traditions had just as much wisdom to offer when it came to questions of mental health and healing. His role at the University of Saskatchewan was to introduce students to Indigenous perspectives in clinical psychology, and one of the things he stressed was that Indigenous healing practices were rooted in synergy—the idea that the whole is greater than the sum of its parts.

Synergy, Katz argued, is foundational to cultures around the world. But with the advent of a capitalist market economy and the seismic ruptures that had accompanied the process of colonization and modernization, a scarcity paradigm based on individualist competition eclipsed synergy as the major driving force of social organization. The dramatically transformed cultures that emerged out of modernity tended to privilege and reward individual rather than collective effort. The guiding metaphors for how society worked became mechanistic rather than organic. Rather than being cells in an organism, for example, workers in a factory were imagined to be similar to the cogs in the machines they tended: more-or-less identical, and easily

P. Cornish (✉)
Student Mental Health, University of California, Berkeley, Berkeley, CA, USA
e-mail: peter.cornish@steppedcaresolutions.com

P. Cornish, G. Berry (eds.), *Synergy*,
https://doi.org/10.1007/978-3-031-88493-1_1

replaced. Through colonial expansion and industrialization, European powers created a world that was simultaneously more uniform and more atomized: Even as individuality was elevated to the primary principle of social organization, actual individuals were expected to conform ever more closely to established norms of the dominant culture. While synergy still operated in some areas—team sports, the arts, community activism—bureaucracies (including healthcare bureaucracies) were organized to function well regardless of whether the people inside them were working together synergistically or not. Nor were countries with communist or state-socialist governments much different: While they were not governed by the profit motive to the same extent, bureaucracies still operated in a top-down way, with little regard for the needs and input of the general population. In both cases, bureaucracies were often designed specifically to thwart the possibility of synergy, insofar as synergy had the potential to be a destabilizing force. There is nothing a bureaucracy fears so much as instability. The zero-sum game of capital accumulation and cultural hegemony are easier to manage and govern by maintaining a culture of winners and losers.

Even in the face of violent colonial projects, Indigenous civilizations, on the other hand, preserved a more synergistic understanding of the world. In his book *Indigenous Healing Psychology*, Katz described what this synergistic understanding looked like in the context of a Fijian village where he lived for several years.

> Synergy occurs when apparently conflicting elements come together, in unexpected patterns, to form a harmonious whole… When synergy is released, a synergistic community is created. In a synergistic community, what is good for one is good for all; in the Fijian fish drive, each community member's collaborative participation not only feeds himself but also feeds every other member… As a result of the entire village's participation, a fish drive can exist, an event whose effectiveness goes beyond a mere adding up of the efforts of each member of the drive (Katz, 2017, p. 349).

A synergistic community recognizes that individuals cannot be completely fulfilled on their own, but reach fulfillment through their cooperation and collaboration with others. The fish drive requires everyone to pool their individual abilities, but the result is a greater harvest than would have been possible had each individual been competing with everyone else. This collective approach is more efficient, but it requires people to trust that they will receive their fair share of the catch. Synergy is the process by which people develop trust—the trust that allows them to work together for the greater good.

As a psychologist, Katz was particularly interested in the way synergy applied to mental health. During his time as a researcher living among the Ju/'hoansi of the Kalahari Desert in Botswana, he studied under = Oma Djo, a Ju/'hoan healer. He learned about their practice of releasing healing energy of *n/om*, "the spiritual energy that permeates Ju/'hoan life, making healing and community development possible" (Katz, 2017, p. 3). For = Oma Djo and the Ju/'hoansi, *n/om* is something that must be handled very carefully, and the process of becoming a healer is about "boiling" the *n/om* in such a way that it can be channeled toward healing (Katz, 2017, p. 96). But it is neither limited nor rationed: As it begins to boil, it flows through the dancers until there is enough for everyone. Having come out of a

Western healthcare context where the value of a treatment is often tied to how exclusive and scarce it is, Katz was struck by the profound implications of the Ju/'hoansi approach to healing. "The Ju/'hoan healing dance, with its renewable, expanding, and accessible n/om," he writes, "is a paradigm both for Ju/'hoan healing and more generally for a healing community" (Katz, 2017, p. 332). The power of *n/om* was a central aspect of life, and it was something that required training and maturity to understand, but it was also something that increased the more it was accessed. What if the Ju/hoansi understood something about healing that had been lost in the West?

The term "synergy" was not one that the Ju/'hoansi or Fijians used to describe what they were doing—they had words in their own languages for talking about their approach to healing and life. Katz borrowed the term from Buckminster Fuller because he believed it was the closest word English had for the kind of collaboration he was observing (Katz, 2017, p. 349). And while he would write about synergy in many of his books, he stressed that it was not an idea he had come up with: He was simply taking a concept he had been taught by Indigenous healers—"the *First Psychologists*," as he called them—and trying to put it in terms that a modern, Western audience could understand (Katz, 2017, p. 5). Synergy was the process by which a group of individuals, sometimes with opposing goals and different mindsets, came together *as individuals*, and, through a process of mutual trust and shared commitment to the common good, create something that is collectively larger than any of the individual components.

This was true in more than just an interpersonal way. Entire societies could operate synergistically when their organization was modeled on the working of the human body, where each distinct feature was acknowledged and its interconnectedness with every other feature was respected. Put in these terms, synergy is fairly easy to understand. But the practice of synergy is, as we shall see in this book, a much more complicated thing.

Synergy and Stepped Care 2.0

The book you hold in your hands (or which you are reading on your screen) is the third in a series that began with *Stepped Care 2.0: A Paradigm Shift in Mental Health* and continued with *Stepped Care 2.0: The Power of Conundrums* (Cornish, 2020; Cornish & Berry, 2023). In the first book, I argued that we need a mental healthcare system that could actually meet the needs of all the people who require mental health support and put forward a model called Stepped Care 2.0 (SC2.0) that would do this by empowering clients to tap into a much wider network of care. As I saw it, one of the reasons mental healthcare was in such a perpetual state of crisis was due to the over-professionalization of the industry. While committed professionals generally seek to do their best, I felt it was necessary to honestly evaluate whether the current system was doing everything it could to achieve the desired outcome—well-being everywhere. This meant asking whether weekly, intensive talk therapy was the best way to meet the massive demand for counseling services, and whether viewing healing as a scarce resource that needed to be doled out by professionals would actually help us fit the resources at our disposal to those with the greatest needs. By expanding the range of treatment options to include self-directed care, peer support, and other forms of community care, I had hoped that SC2.0 might help make mental healthcare more efficient, more effective, and more empowering.

A lot has happened since *A Paradigm Shift in Mental Health* came out. There was a global pandemic, a cost-of-living crisis, and a housing crisis, and all the while, the climate emergency has worsened, and violent conflicts have spread in many parts of the world. The need for effective, empowering care is as acute as it has ever been. However, while SC2.0 spread to provinces, territories, and universities across Canada and the United States, we encountered many challenges with this work. The second book in the series, *The Power of Conundrums*, consisted of a series of reflections from professionals who have studied, implemented, or worked within the SC2.0 model on the practical difficulties involved in bringing the model to life. We referred to these difficulties as "conundrums" because we wanted to view them not simply as obstacles, but as sources of creative inspiration. What was the SC2.0 model missing? Which forms of resistance on the part of clients and practitioners had we failed to foresee? When SC2.0 did not achieve successful implementation, what were the root causes? What did successful implementation even *mean*, given that we were trying to reform how mental healthcare is conceived and delivered dramatically? Had SC2.0 just become another framework among many, or was there still a way to transform it into a movement? The conundrums did their job by getting us to think more deeply about the limitations of SC2.0 and the deeper structural and philosophical problems that exist within Western systems for mental and physical care. This thinking has led us to a bold conclusion: If SC2.0 was simply a model for delivering care that did not take into account the particular gifts, limitations, and capacities of the people implementing and delivering it, it would not fundamentally transform the system. In order to properly function, the kind of synergy Katz observed in Indigenous communities around the world was required. If SC2.0 was the motor that could propel us in the right direction, synergy was the electricity that gave it power.

In this final volume, we will explore synergy in greater depth and provide an account of what synergistic implementation of SC2.0 can look like. Because synergy is an idea that is philosophical in nature, this volume draws on a number of critical tools from a wide range of disciplines in both the humanities and the social sciences. Understanding the literal meaning of synergy is deceptively simple, but to grasp how it can be used to transform our existing systems of care, it is necessary to understand its long history and deep roots in non-Western healing traditions around the globe. For this reason, we have chosen to make storytelling a key part of this text. So, it is only appropriate that when describing how synergy relates to SC2.0, I start with my own story.

Until very recently, I did not really know why I was doing the work that I was doing. At a European congress in 2024 with the subtitle "Psychiatry in Crisis," I was enthralled by the depths of insight, the philosophical musings, the artistic expression, and the powers of reflexive discourse offered by the mix of speakers. All participants began with and centered their presentations on lived expertise. One psychiatrist's presentation consisted only of entries from a personal diary for a year as he descended into and recovered from a deep, psychotic depression. An artist with no formal psychiatric training represented her experiences drifting in and out of psychosis as both a painful yet wonderful expression of what it is to be a variant of normal. She guided us through a series of self-portraits, drawing several times on contemporary philosophers to note the mindsets, assumptions, and structures that had previously dehumanized her experiences.

I had prepared a basic description of SC2.0 for my presentation, but I decided to abandon the PowerPoint in favor of an in-the-moment reflection on how I came to do the work I do. I shared a realization that who I was and the people who accompanied me through the first part of my life were much more than I had thought. My father was not a man diminished by bipolar one disease. He was a bright, highly energized, fun-loving person, a beloved character who needed respect and accommodations during his long, dark bedridden days. His brother, my uncle, was not the overly medicated paranoid schizophrenic, nodding off in his recliner, occasionally drooling. He was a gifted pencil and paper artist, representing irony and lightheartedness in the figures he scratched on small scraps of paper. At the same time, at age five, I sat on the floor making funny faces just to hear his guttural, somehow deeply infectious laugh. My sister, born 3 years premature, with a mild cognitive delay, multiple learning disabilities, and a late-in-life diagnosis of schizophrenia, is the keeper of common sense in our family. Both my brother and my mother, functioning alcoholics, introverts, and gifted teachers, share a passion for a deep connection to nature and the land we grew up on.

At the congress, I shared their stories, their lives lived in conflict with inner demons. I shared their unique gifts. I shared how it was easy to judge them and feel embarrassed by their failings. That is, until I slowly began to realize that the failing was on my part, on that of my profession, and the colleagues who had come before me with all the best intentions. I had been equipped with blinders, seeing my family as riddled with deficits and defects. But at the congress, I witnessed with awe the yin and the yang of experience in powerfully dramatic stories in the presentations before me. I saw something new. I saw the conflicting parts of my family experience as dynamic, rich, fully human, and grounded in connection, love, and the land that had nurtured us. My epiphany was evidence for me that I was tapping into synergy.

As mentioned above, I was first introduced to the idea of synergy while working on my master's degree. I had entered the program with the intention of studying the self-help movements of the 1980s because I was interested in the way they sought to create a language of agency people could use to empower themselves. The experiences of my family members revealed to me how the psychiatric industry had obscured their unique gifts and strengths by pathologizing their challenges. Even then, I was concerned with the industry's reliance on professionals, and wanted to explore forms of care that relied less on the established practitioner–client dyad. People living with mental illness—and especially those experiencing intersecting forms of marginalization—are among the most systematically disempowered in our society, and it seemed that the mental healthcare structures only further entrenched this experience of disempowerment. But as I went deeper into the literature on empowerment, I began to feel something was missing. The concept was defined in many different ways, but it was generally assumed that empowerment was about autonomy, assertiveness, and having or taking control. This struck me as being a highly individualistic understanding of the phenomenon, one that ignored the collective dimension of empowerment that is central to political activism and community-building. Was there a language for empowerment that was not simply about maximizing individual agency, but which acknowledged that individual well-being often requires a sense of belonging to something larger than oneself?

These questions spurred me to engage more deeply with Katz's work on synergy. Synergy seemed to provide a complementary schema for understanding how the individual and communal aspects of well-being fit together, one that might even be applied in modern healthcare institutions. Instead of imagining that we as individuals need to give up a measure of our own well-being for the greater good of the community (and in the zero-sum individualism of the West, this is often how the relationship between the individual and the group is conceptualized), synergy helped me understand that the individual's empowerment and well-being actually relied on their ability to collaborate with others in the context of a community, institution, or workplace an open, honest, and vulnerable way. I applied this insight in my dissertation to questions of allyship as they related to men who identified as feminists seeking to cede power to their women colleagues. Rather than seeing power as something that could only be held *over* another, synergy opened up the possibility of having power *with* others—which is to say, conceptualizing how a group can be collectively stronger and healthier if they stop seeing power as something that must be guarded and hoarded, but as something that can be multiplied through sharing.

Synergy helped to shape my worldview as a student. However, when I entered the clinical world, I was quickly overcome by other concerns and pulled back down to the reality of working in alignment with well-established standards of professional practice, including the usual hierarchies of education, credentials, and institutional responsibilities. It was not until I landed a leadership position at the student counseling center at Memorial University in Newfoundland that I began to revisit some of these ideas. As an administrator, I was quickly confronted with the scarcity paradigm that governs most modern institutions. My professional life was defined by limited budgets and unlimited needs. I did not have a large number of staff, and the staff I did have were not operating as effectively as I would have liked—and yet the needs of the students were overwhelmingly pressing and urgent, as were the needs of the communities beyond campus (many of which were remote and lacked on-site psychologists, social workers, and psychiatrists). I knew I would not receive the resources I needed through institutional channels, and the staff of the counseling center were unprepared to change how they worked. So, what was I to do?

I had not yet started to consciously think about this problem in terms of synergy, but my previous work had underscored for me that mental healthcare does not just happen in the clinic. As I began to reach out to other community stakeholders interested in improving mental health outcomes in Newfoundland and Labrador, I started to realize that creative solutions were possible—if we were willing to move beyond our professional silos, meet clients where they were, and start identifying the informal social supports, like knitting circles, and local institutions, like libraries and churches, that already existed in these communities, and which were already practically involved in providing various forms of care to community members. This was the first step toward the realization that these communities did not represent "problems" that needed to be solved; they already contained healing energy, and our task was not to dictate the terms by which that healing should happen, but to offer ourselves as partners in the process. This led me to develop what would eventually be called SC2.0, a story that is covered in greater depth in the first volume of this series.

Dick's mentorship early in my career was formative, and set me on a lifelong path in search of synergy. Twenty-five years later, I met Gillian Berry, who reached out to me for guidance on implementing SC2.0 at The George Washington University. Though I was brought in as a consultant, Gillian's gifts as a teacher meant that the relationship quickly became one in which we both learned from each other. In contrast to a unidirectional formative mentor relationship, our work became reparative, restorative, and symbiotic. Gillian valued my subject matter expertise and my capacity to assist in what we later called code-switching (McCluney et al., 2021), translating for White audiences her cultural wisdom and the sometimes blunt but always loving ways of communicating typical of her Guyanese or assumed "West African" heritage. I benefited from her insistence that I develop and disseminate the ideas I was developing, arguing that what I was doing was revolutionary and needed to reach a wider audience. I was not so sure, but she persisted, and eventually, we published an article titled "Meeting the mental health needs of today's college student: Reinventing services through Stepped Care 2.0" together in 2017. "Now you have to write a book," she said.

But actually writing the book seemed more difficult. The implementation at both Memorial and The George Washington Universities had met with initial success, but had not been implemented as fully as I would have liked, and these partial failures were painful. It was hard not to doubt ourselves. Ironically, while I was filled with doubt about my own abilities, I was able to help Gillian see that her own work (which had, similarly, been initially successful) had stalled due to much larger political and economic factors beyond her control. She helped me see the larger irony of this situation: We could both see what the other was missing. We concluded from this that we had indeed taken a big step toward something, and we set out developing additional parts that would expand the impact of SC2.0. For me, this led to the writing of *A Paradigm Shift in Mental Health*, and shortly after its publication, I had the opportunity to take a position as the Director of Student Mental Health at the University of California, Berkeley, which was implementing SC2.0. As a large, affluent, elite university in one of the richest regions in the world, Berkeley was in a very different situation from Memorial University. But in other ways, it was facing the same challenge: How to make the available resources cover all of the existing needs. At Berkeley, I met many passionate individuals who were committed to healing, and saw things like mental health ally training being practiced. I realized that we could not just run a clinic that served people. We needed to be a clinic that taught people how to serve others. There is, after all, something passive and disempowering about the experience of being served by the mental healthcare system, and often what clients who are struggling need is a sense of purpose. One of the best ways to heal yourself is to help someone else who needs healing.

The mental healthcare systems I had spent my adult life working in did not appreciate this crucial aspect of healing. The mainstream, individualistic approaches emphasized the importance of changing your thinking, removing your deficits, and being more realistic about your situation. What they were missing was language that could address the spiritual parts of ourselves, the parts that yearn for meaning and mission. When I started to see that our job was to empower the community to take care of itself, it gave me as an administrator a sense of hope that maybe this job could be done—this impossible job of serving an ever-larger group with ever-greater needs who wanted a service we had a limited amount of.

Reflecting on the issues raised in *The Power of Conundrums*, my colleagues at Stepped Care Solutions (SCS) and I have noticed that many of the challenges we have faced in implementing SC2.0 have come down to a failure to understand this missional and even spiritual dimension of the model. Stepped Care 2.0 is not meant to be a model that merely provides administrators and executives with a wider range of options for providing mental healthcare in a centralized, bureaucratic, top-down way. It is meant to be a collaborative model in which the healing power that is already latent in communities is unleashed and coordinated in a way that takes the pressure off intensive options (therapists, psychologists, and psychiatric hospitals) by empowering ordinary people to heal themselves and each other. All too often, organizations seeking to implement SC2.0 have been attracted by the promise of reduced wait times and more efficient clinics, but have not been willing to countenance the more radical aspects of the program because doing so would destabilize the hierarchical structures in which they operate.

One of the primary examples of this is the attitude toward One-at-a-Time (OAAT) counseling. This approach to treatment is based on the fact that many people will only come for a single session of therapy, and so it treats each therapy session as if it will be the only one.[1] Because OAAT allows more clients to be seen more quickly, it is an appealing intervention for clinics that are dealing with massive backlogs and clients whose urgent needs are not being met. But OAAT is not just a new administrative model: It also represents a way of thinking about client care that acknowledges the abundant healing resources in the wider community, and views the clinician's role as being partially about helping the client connect with these resources—be they community groups, peer support networks, religious institutions, or self-directed activities. Naturally, clients who need more intensive supports, such as medication or in-patient treatment, will be connected to these

[1] A common misconception about OAAT is that it *only* provides one session. In practice, clients may seek as many sessions as they need, but they will do it on a session-by-session basis.

resources and care providers. But the vast majority of people walking into clinics or seeking out therapy do not need to speak with a licensed professional every week. What they need is attention, empathy, validation, connection, and a listening ear, which could be provided in a much less resource-intensive way through networks of support that may already exist in their community or online.

Connecting people to these networks, rather than signing them up for a multi-session course of treatment, does not just help reduce the pressure on systems of care so they can direct resources toward more acute cases. It also helps us shift our attitude toward care from being one in which healing is something a professional bestows on a client to one in which healing is something the client does in the context of a much larger community. One of the most profound things about peer support systems like Alcoholics Anonymous is the way they turn each participant into a partner in healing. A person who moves through the 12 steps benefits from the help and support of their sponsor, but they also help the sponsor reaffirm their own commitment to sobriety; in time, they may become a sponsor themselves. This is synergy in action: The healing is passed along between people, and the further along their own healing journey a person travels, the more healing energy they can pass along to others.

This kind of synergy is the fuel that powers SC2.0. What we mean by this is that the model and structure of SC2.0 is an engine that can help connect the disparate parts of our networks of care (both inside and beyond the formal healthcare system). But the engine cannot run without passion, energy, and the individual capacities each individual within the system brings. To return to the example of peer support, AA could not afford to operate if every sponsor needed to draw a salary, and every AA group needed to set up complex intake processes, purchase office space for its administrators, and hire permanent legal staff. Like so many community groups, AA runs on the dedication its members have for the mission. One of the most remarkable things about any community effort is the degree to which people participate simply because they see intrinsic value in what they are doing. The SC2.0 model functions well when it recognizes that these are the true agents of support, well-being, and change, and that the role of the mental health administrator should be, in large part, a matter of facilitating the tools for healing that already exist in a community by unlocking the synergistic capacities that are latent within it.

Stepped Care 2.0 grew organically out of an attempt to unblock the bottlenecks that so often reduce how much support is available—not by pouring more resources into an industry that simply cannot meet the scale of demand on its own, but by recognizing that a group of people working together synergistically can provide a truly surprising amount of healing. Our role as mental health professionals should not simply be to serve those who are unwell: It should be to teach people how they can serve others (including themselves), so that each person who receives care is also empowered to spread that healing to the people around them. This is the power of synergy, and it is why SC2.0, without synergy, is simply another tool for inefficient administration.

The Evidence for Synergy

One key conundrum addressed in the previous volume dealt with the question of evidence: Namely, whether or not there was any scientific evidence showing that stepped care models like SC2.0 led to a measurable increase in the quality of care for key demographics. In that chapter, authors Amy Salmon, Mai Berger, Saranee Fernando, and M. Elizabeth Snow concluded that the lack of a consensus definition for what "stepped care" refers to (given the wide array of models that describe themselves as providing "stepped care") made it difficult to judge the case either way. The chapter pointed to one of the key methodological challenges facing the social sciences: In order to be considered legitimate, a treatment must be backed by evidence that it has an effect, but our tools for measuring impact and identifying its cause are not epistemically neutral (Salmon et al., 2023).

The scientific method excels when dealing with controllable, repeatable phenomena, but the phenomena we encounter in psychology—especially on the population level—are not always controllable or repeatable. This means the interventions with the largest evidence base are not necessarily the most effective—they are just the ones whose effects are easiest to measure.

An intervention like SC2.0, which is meant to be adapted to local conditions and which deals with complex group dynamics, is much harder to test than Cognitive Behavioral Therapy. But this is not the only thing that makes gathering evidence for SC2.0 difficult. Because SC2.0 is intended to transform mental healthcare, it has an activist, missional edge, and trying to measure whether it will have a net positive impact before it is implemented is a little like trying to develop a metric to determine whether a particular political party will preside over a successful administration. There are so many unknowns, so many variables that are subject to change, that the decision needs to be made based not only on the specifics of any one particular policy, but on the party's vision for the good of the country. If we like a particular politician or political party, it is generally because we find ourselves aligned with their sense of what the "right thing" looks like, and trust that they will respond to unforeseen challenges in a way that is responsible and compassionate. This is why Alexia Jaouich, Stacie Carey, Kaitlin Saxton, Joshua Rash, and Jacquie Brown argued in the previous volume that the "active ingredients" of SC2.0 were the model's nine core components[2] and the 10 guiding principles (Jaouich et al.,

[2] The Core Components of SC2.0 are: (1) Systems are co-designed with people who bring diverse perspectives and experiences. (2) Systems reflect a variety of step levels and include both formal and informal services. (3) The management of risk is distributed across the systems.

(4) System and service improvement is achieved through continuous monitoring and quality improvement cycles. (5) Recovery-oriented approaches are clearly and consistently integrated at all levels of the systems of care. (6) People have same-day access to multiple levels of care. (7) Resources and services are guided by a one-at-a-time approach. (8) Services are flexible, data-informed, and collaborative. (9) Care is person-centric.

2023).[3] SC2.0 is a way of *thinking* about mental healthcare as much as it is a model for delivering it, and if an organization that implements SC2.0 is fully aligned with the core components and guiding principles, the results will be consonant with the mission SC2.0 seeks to fulfill.

If it is hard to provide scientific evidence for the effectiveness of SC2.0, it is even more so for synergy. This is because the roots of synergy lie not in social science, but in philosophy—and Indigenous philosophy in particular. Studying synergy properly requires us to look beyond the methodological tools offered by psychology alone. Setting out to write this book, we decided to incorporate traditional social science tools like surveys while also drawing on philosophy, anthropology, and journalism. In particular, we sought to incorporate the anthropologist Clifford Geertz's method of thick description (Geertz, 1973) in describing situations that provided examples of synergy in action. But we found that when it came to capturing what synergy looked like in practice, the most useful tool of all was storytelling. Describing what synergy is and explaining how it works cannot really be done without telling a story about what happened—how it felt like to participate in a synergistic process, for example, or why you felt more respected and productive working as part of one team compared to another. While this is an unorthodox approach for social scientists to take, we believe unorthodoxy is needed in order to capture what works about synergy.

There are three reasons for this. The first is that storytelling is experiential. When storytelling is done well, it helps us understand that person's perspective and behavior in ways that can help us make sense of broader phenomena, and it can help us question some of what we think we know. Jeff Bezos, founder of Amazon, once said that when the anecdotes conflicted with the data, he trusted the anecdotes, because they were often a sign of problems the data was missing (StoryWorks, 2024). As psychologists, we can sometimes fall into the trap of thinking that we know better than our clients, and that if they simply did what we told them to do, they would be able to heal. But sometimes the data (or the theory) is wrong, and the anecdote is right. This is why we need to listen to what people say, even—or perhaps especially—when what they say conflicts with our deeply held professional view of things. Given how many people (and especially BIPOC and gender nonconforming

[3] The Guiding Principles of SC2.0 are:

- Social justice drives effective care systems transformation and is an intervention in itself.
- Multiple and diverse care options are required as one approach will not work for everyone.
- All individuals and communities have strength and capacity.
- People engage with what they are ready to do; gold standard intervention is that which best fits the service user at any given time.
- Professionals do not carry all the wisdom; people often know what is best for them.
- Mental health literacy is required for people to make informed decisions.
- An effective care system ensures people have access to care when and where it is needed.
- The whole is greater than the sum of its parts: The strength of the system relies on multilevel collaboration.
- Minimal interventions can produce powerful results.
- There is no ideal solution; trial-and-error leads to growth and change.

people) report feeling alienated by or unsafe within the bureaucratic system of mental healthcare, perhaps it is worth asking what their stories tell us about the shortcomings of our way of working.

The second reason why storytelling is an appropriate way to provide evidence for synergy is that storytelling provides vital context that can be difficult to capture without it. In the sixth chapter of this book, Dr. Gillian Berry describes how studying with Kitso Mbenge, a traditional healer working in Botswana, helped her to understand some of the limitations of her own training. In order to do so, she provides an account of her own upbringing and how her experience as a Black woman in the British education system led her to become critical of some aspects of the Western approach to healing. The synergistic healing she observed in Botswana (which reminded her of her own grandfather's healing practice in Guyana) was in stark contrast to the managerial and often impersonal approach to healing that she had become disillusioned with in Britain. This personal context may be difficult to measure quantitatively, but it is essential for understanding the arguments she makes about the role traditional wisdom can play in modern healthcare systems.

The third reason storytelling is important is that it cannot be avoided. Without a story, data is just numbers, and one of the shortcomings of an approach to knowledge generation that views numbers as the be-all and end-all is that they are susceptible to unconscious bias—we may want to view the numbers in light of the story we want to believe is true. Data is valuable because it helps us gain an understanding of what is happening, but data is only useful if we are able to build an accurate account of *why* it is happening. And as soon as you are talking about "why," you are talking about stories. By unapologetically basing this book on narratives, we acknowledge that each individual's experience is a form of data, and that creating a larger account of what these stories mean is, to some degree, a creative, storytelling act. We understand that this method is atypical for a book about mental health systems, but we believe that to grasp a concept as fundamental as synergy, it is necessary to consciously and critically embrace the inevitability of storytelling as a way of understanding the world.

Finally, we believe that writing a book about synergy can only be done synergistically, which is why the chapters that follow include a wide range of different authors and perspectives. Some are mental health industry veterans, others are early career professionals, and others still have no specialist knowledge in psychology, but have agreed to participate in the book as outside observers. Not all of the authors, or the people they interview, are in complete agreement about every aspect of synergy. But as Katz has argued, synergy itself is about the coming together of apparently conflicting elements, and we hope that by providing a multiplicity of voices from different countries, backgrounds, demographic groups, and professional spheres, we can provide a well-rounded account of synergy's importance, and why it is essential to the functioning of SC2.0.

Summary of Contents

Unlike some academic books that present the work of several individual researchers on a particular topic, the chapters in this book are meant to be read in sequence. This introductory chapter sets the context for what synergy is and why it matters. In the next chapter, I return to the conundrums about SC2.0 raised in the previous volume, and make a case-by-case argument for how synergy can be used to address them. From co-design to the risk paradigm, the conundrums we explored each point to the ways in which the SC2.0 model on its own is not enough. In order for SC2.0 to deliver on its promise to transform mental healthcare, it relies on the power of synergy to break down the barriers—between clinicians and clients, administrators and staff, professionals and volunteers—that keep us from achieving well-being everywhere.

The third chapter, authored by André Forget and based on interviews with Genée Jackson, Raghav Suri, and Kareem South, looks at why modern institutions are so often resistant to synergy, and how each of these individuals has tried to work synergistically in institutional contexts. In the fourth chapter, "The Gaslighting of a Profession: DEI Training in the Mental Health Profession," Catie Greene picks up these questions in the context of Diversity, Equity, and Inclusion (DEI) training as it relates to mental health practice. While DEI was originally formulated as a way of confronting the systemic inequalities and biases present in Western institutions (including ongoing practices of racism, colonialism, misogyny, heteronormativity, queerphobia, and ableism), it has increasingly functioned as a bureaucratic mechanism for deflecting criticisms away from institutions without fundamentally addressing the issues at play. Greene argues that synergy shares many of the same goals as DEI, but is committed to a more radical practice of inter-subjectivity in which the fullness of each person's identity can be acknowledged in the process of building a truly anti-racist and decolonial system of care.

The fifth chapter, "A Journey of Discovery," follows the journalist and novelist André Forget's experience over the course of a week spent observing the filming of a documentary about the work of Dick Katz. The documentary was shot outside Parry Sound in October 2023, and brought together Indigenous Elders, academics, psychotherapists, and employees of SCS interested in the concept of synergy. As an outsider to the mental health world, Forget brings a critical eye to the challenges involved in building synergy among this diverse group of participants. Through his conversations with the documentary crew, SCS leadership, visiting professors, and the local people involved in putting the shoot together, he explores the power and challenges of working synergistically.

Berry's "The Education of a Healer" tells the story of how she moved to Botswana to study traditional healing practices, and how the insights she gained from the traditional healer Kitso Mbenge influenced her own practice as a social worker in the United States. By exploring the training and formation of healers in Botswana, which focuses on spiritual calling, community input, and individual character, Berry draws attention to the way Western approaches to training healers turn this important social function into a job like any other. Berry argues that the training healers like Mbenge undergo is both communitarian *and* more respectful of individual strengths and capacities, while Western systems (which only judge candidates on their willingness and their academic abilities) do not require input from the community the trainer will serve, and flatten out what makes each individual candidate unique. The chapter brings together personal history and Afrocentric philosophy to provide a profile of Mbenge, an analysis of traditional healing practices, and a practical guide to how traditional wisdom (and synergy in particular) can be applied in the context of a modern mental health bureaucracy.

In the penultimate chapter, "Case Study: Building Synergistic Systems of Mental Healthcare," Sarah Harte picks up where Berry left off, with a detailed analysis of how synergy can work in the context of the mental health clinic of a large US university. Drawing on data from a series of surveys conducted in 2022, and her own experience working at the clinic, Harte explores the role synergy played in the implementation of SC2.0. Harte argues that the implementation was only possible because a group of key clinicians and stakeholders worked together synergistically to explore a bold alternative to business as usual. Through their commitment to engagement with the clients, student body, clinical staff, and university administration, they were collectively able to implement a new model for care that had a profound impact on the lives of both staff and students. Harte's chapter ends by highlighting one of the main challenges facing any synergistic way of working: its vulnerability in the face of hierarchical, authoritarian styles of leadership that are threatened by synergy in action.

Conclusion

The first book in this series began by acknowledging that mental healthcare is in crisis. While many proclaim that we have a mental health crisis (that is, that many more people are suffering from symptoms of mental illness), we shifted the focus to examine *structures* that are in crisis. But while many of the solutions being discussed in the mental health industry today—including more e-mental health options, AI tools, and the exciting potential of psychedelic treatments—certainly fit within the framework of expanded options explored in *A Paradigm Shift in Mental Health*, I am not sure they will be sufficient for two reasons:

1. The malignancies that led to this crisis of access are embedded deep within the system. As long as we cling to a care model that privileges access to highly paid professionals and treats the ancient healing practices humans have used for millennia as unscientific and unworthy of consideration, the crisis will continue.
2. We are in the midst of a myriad of socioeconomic crises. The climate emergency is ever more present. Social and spiritual disconnection is on the rise, as is disconnection from the natural world around us. Binary left/right political polarization is on the rise, as is distrust of all forms of leadership. To frame this as a

> mental health crisis would be akin to describing war as an epidemic of premature mortality. While recruiting and assigning more medics and surgeons to war zones is a necessity, it will not solve the problem.

With SC2.0, we have articulated what a better approach might look like. However, experience has taught us that if SC2.0 is simply slotted into a system that persists in the same hierarchical, Eurocentric ways of organizing labor and thinking about healing, it will not change much. We need to learn from the First Psychologists and start taking synergy seriously. These ideas about synergy do not just apply to mental healthcare and healing: Every industry and workplace could benefit from becoming more synergistic.

As you will see in this book, we do not treat synergy as a panacea, some kind of miracle drug that can cure the healthcare system overnight. Synergy relies on people's willingness to work together, not in spite of their differences but *through* their differences, respecting the unique perspective that each individual brings. In a period of global turmoil, rising inequality, and a catastrophic decline in trust, this is not an easy fix; it is an extremely hard fix, one that requires us to work through our fears and anxieties. Building synergy in the context of capitalist bureaucracies founded on principles of scarcity, managerialism, and profit is profoundly countercultural. Nor is synergy something that can be established once or for others. Instead, it is continuous and collaborative. It is something that must be created anew every time we sit down with a client, every time we hold a budget meeting, because it relies on an ongoing willingness to see the people around us as partners to engage with, rather than "problems" to manage. We believe that the stories in this book will show that the struggle *with* others is worth it.

It is important to note that this book was completed prior to the escalation of political and social upheavals in early 2025. Rising political and economic turmoil, shifts in immigration policies, and the growing debate over Diversity, Equity, and Inclusion (DEI) initiatives are adding stress to people worldwide, and further straining already-inadequate health and healing systems. The resulting instability, including deportations and the politicization of social movements, profoundly impacts marginalized communities, and directly challenges the frameworks through which we provide care and support. Amid increasing polarization and divisiveness in the political sphere, there is a pressing need for a paradigm shift in mental health systems. This shift must prioritize the decolonization of care, and empower individuals and communities to heal amidst political turmoil. It necessitates a synergistic, holistic, and collaborative approach to mental health—one that embraces core values such as truth and reconciliation, interconnectedness, empathy, and flexibility. These values are essential for navigating these times of intense social and political division. They require crossing the aisle with an openness to understand political perspectives and conflicting values that have been disconnected or dismissed. The principles outlined in this book are more relevant than ever, and they call for a deep understanding of and learning from society's diverse perspectives and reactions. This inclusive approach is critical for fostering collective healing and resilience across all communities.

A final note: I was trained as a psychologist and researcher, and I have spent my career working as a leader in the world of mental health. It is work that I am passionately committed to. But in the process of putting together this trilogy of books, I have come to realize that being a leader of an organization that takes synergy seriously means that I need to see myself not as the soloist, but as one among many voices in the choir. I would like to think that the progression of this series of books illustrates that point. I was the sole author of the first volume, and the convenor and editor of the second. This final volume, with its many viewpoints, personal reflections, and stories, is a further departure from the standard academic text. It has evolved considerably over the course of writing, as each author has shared their own vision for what the book might look like. It is a collaborative effort, and we hope that readers will appreciate that we have all contributed, in one way or another, to each of the pieces that make up its contents—while allowing the author's unique voice to come through. It is, we hope, not just a book *about* synergy, but a book that emerged *through* synergy.

References

Cornish, P. (2020). *Stepped care 2.0: A paradigm shift in mental health*. Springer. https://link.springer.com/book/10.1007/978-3-030-48055-4.

Cornish, P., & Berry, G. (Eds.). (2023). *Stepped care 2.0: The power of conundrums*. Springer. https://link.springer.com/book/10.1007/978-3-031-45206-2.

Cornish, P., Berry, G., Benton, S., Barros-Gomes, P., Johnson, D., Ginsburg, R., Whelan, B., Fawcett, E., & Romano, V. (2017). Meeting the mental health needs of today's college student: Reinventing services through Stepped Care 2.0. *Psychological Services, 14*(4), 428–442. https://doi.org/10.1037/ser0000158

Geertz, C. (1973). *The interpretation of cultures*. Basic Books. https://philpapers.org/archive/GEETTD.pdf.

Jaouich, A., Carey, S., Saxton, K., Rash, J., & Brown, J. (2023). Adaptability and fidelity: Exploring the relationship between the two when implementing SC2.0. In P. Cornish & G. Berry (Eds.), *Stepped care 2.0: The power of conundrums*. Springer. https://doi.org/10.1007/978-3-031-45206-2_11

Katz, R. (2017). *Indigenous healing psychology: Honoring the wisdom of the first peoples*. Healing Arts Press.

McCluney, C. L., Durkee, M. I., Smith, R. E., Robotham, K. J., & Lee, S. S.-L. (2021). To be, or not to be…Black: The effects of racial codeswitching on perceived professionalism in the workplace. *Journal of Experimental Social Psychology, 97*., Article 104199. https://doi.org/10.1016/j.jesp.2021.104199

Salmon, A., Berger, M., Fernando, S., & Snow, M. E. (2023). Stepped care, a closer look: Reviewing evidence and evaluating outcomes. In P. Cornish & G. Berry (Eds.), *Stepped care 2.0: The power of conundrums*. Springer. https://doi.org/10.1007/978-3-031-45206-2_9

StoryWorks. (2024, July 22). Jeff Bezos—Data versus anecdotes [Video]. YouTube. https://www.youtube.com/watch?v=uFUc_5OMB5s.

Chapter 2
Synergy and the Conundrums of Stepped Care 2.0

Peter Cornish and Gillian Berry

Conundrums and the Strands of Synergy

In the decade since the Stepped Care 2.0 (SC2.0) model was first proposed, it has been implemented in several Canadian provinces and territories, as well as student mental health centers in Canada, Mexico, Chile, and the United States. Interest has grown internationally as well, with collaborators in Finland, Norway, Wales, Singapore, and the Netherlands reaching out to learn more. As SC2.0 expanded, however, we, along with others at Stepped Care Solutions (SCS), noticed patterns in the problems we faced when implementing the model. Its radical potential was not always understood, and we sometimes found ourselves working with partners who only wanted to implement the parts of the model that fit most easily into their established paradigms.

We began to think of these problems not as issues that needed to be put to rest as quickly as possible, but as conundrums that could deepen our thinking about the challenges of building a synergistic model of care. The power of conundrums lies in their capacity to fire our imaginations and engage our creative spirit. If we are willing to sit with them, they can point us toward surprising insights and new revelations. In response to these conundrums, we solicited contributions from other team members, outside scholars, mental health practitioners, and academics who had experience working within the SC2.0 model. Each participant had insights to share about what made the model sometimes challenging to understand and difficult for

P. Cornish (✉)
Student Mental Health, University of California, Berkeley, Berkeley, CA, USA
e-mail: peter.cornish@steppedcaresolutions.com

G. Berry
Stepped Care Solutions, Mount Pearl, Canada
e-mail: gillian.berry@steppedcaresolutions.com

P. Cornish, G. Berry (eds.), *Synergy*,
https://doi.org/10.1007/978-3-031-88493-1_2

some partners to implement. These conundrums were published as *Stepped Care 2.0: The Power of Conundrums* in 2023 (Cornish & Berry, 2023).

Some contributors proposed solutions to the conundrums they investigated, while others focused on identifying tensions, or telling stories of successes and failures. The purpose of the book was not to provide a clear set of easily actionable solutions, and few of the chapters ended with a clear prescription. But having explored the conundrums in a spirit of openness and curiosity, it is now time to address them. In this chapter, we will look at each of the conundrums individually and show how they arise from a failure to work synergistically, before making an argument for how synergy can be applied to each challenge.

Before we begin to address the conundrums themselves, however, we must offer a clearer account of synergy as imagined by SC2.0. While synergy is a phenomenon common to all cultures, we have identified 10 principles, or interwoven "strands," across three planes, that define what synergy looks like. As illustrated in Fig. 2.1., the strands are: Truth and reconciliation, interconnectedness, diplomatic disruption, awareness, creativity, contextualization, growth orientation, flexibility, agency, and empathy. These strands of synergy combine and interact across the planes of personal, interpersonal, and collective responsibility. When synergy is working, each of these strands supports action that is reparative, productive, and empowering for individuals, communities, and social groups. While this is not meant to be a

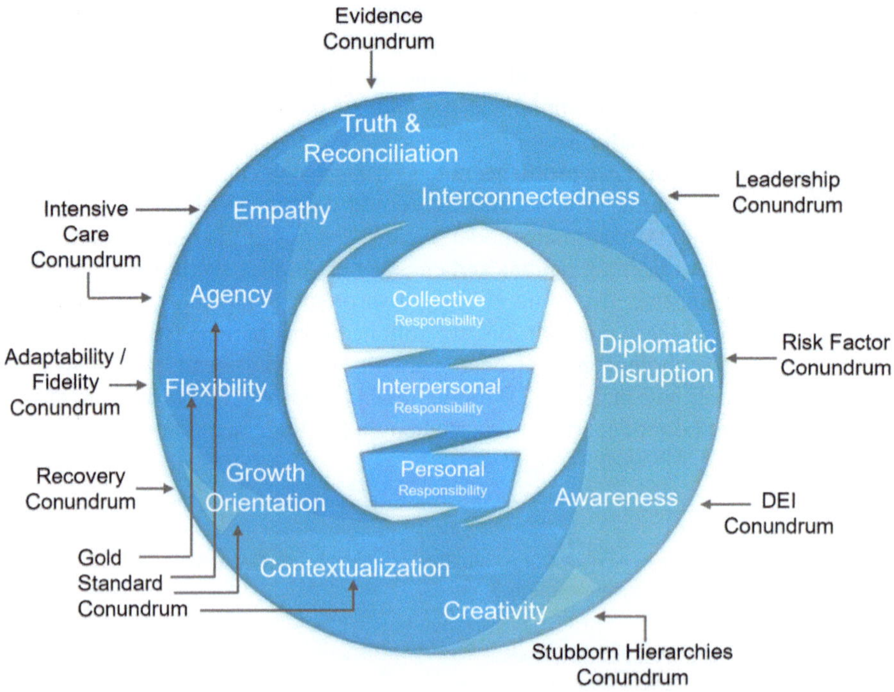

Fig. 2.1 The strands of synergy and their relationship to SC2.0 conundrums

complete account of *everything* that synergy is or could be (it is possible that within a different cultural context from the one in which we are writing, other strands might be added), we have formulated the strands as a guide to help people understand what synergy looks like in practice.

Each of the nine conundrums identified in the previous volume expresses a tension or an energy that becomes productive when connected to the strands of synergy. In Fig. 2.1, each conundrum is paired with one or more strands of synergy that, as you will see in the sections below, can contribute to a more expansive and sustainable healing system. While some strands are more relevant to some conundrums (as indicated by the proximity of a strand label to a conundrum label), it is important to note that synergy works because the strands are interwoven, and do more in combination to channel the tensions productively than they could do alone.

Addressing the Conundrums Through Synergy

Harnessing the Power of the Evidence Conundrum Through Truth and Reconciliation

The first conundrum raised a challenge that had become significant as SC2.0 began to expand: the conundrum of evidence. Administrators, academics, clinicians, and clients considering the new model all wanted to know that it would be effective. Where was the evidence that SC2.0 worked? Amy Salmon, Mai Berger, Saranee Fernando, and M. Elizabeth Snow looked into the published, peer-reviewed literature on "stepped care" and came to the conclusion that while this research was "far behind existing practice," this was not because there was not any evidence for stepped care, but because the phrase "stepped care" had been used to refer to so many different models that no consensus definition of stepped care existed. Furthermore, the tools used to evaluate effectiveness were often limited in scope, and far better suited to measuring the impact of new medications. They concluded that new evaluation resources were needed if the effectiveness of SC2.0 could be proved one way or another.

The evidence conundrum raises a number of important questions about the epistemic biases present in contemporary mental health research. Synergy is an interaction between different agents, organizations, or substances that create a combined effect that is greater than the sum of its parts. Because academic psychology and medical research often seek to isolate the single active ingredient (or ingredients) responsible for an effect, the research tools available tend to be optimized for experiments where every input can be tightly controlled. However, measuring the effectiveness of a healthcare system shaped by millions, if not billions, of inputs using these tools is akin to measuring the aesthetic impact of a painting by analyzing the chemical bonding in the paint. It can produce results, and it can even generate insights, but it is not always a useful guide to understanding whether the painting will have an emotional impact on a viewer.

A synergistic approach does not reject evidence or discount the notion that some systems are more effective than others, just as some treatments are more effective than others. But it understands that quantitative data only ever provides a snapshot, and this snapshot is determined by several human and environmental factors (where was the camera placed? How close or far away from the subject was it? How bright was the light?). A wide range of research tools need to be used in order to gain a more complete picture. Some of these tools may be rooted in the scientific method, while others might come from the social sciences, philosophy, the arts, or even journalism. A synergistic view of evidence accepts that change is often caused by a vast number of factors, some of which might be more important than others, all of which matter.

Truth and reconciliation is the pillar of synergy most relevant to the question of what constitutes evidence. A synergistic approach to validation and evaluation must actively decolonize the research enterprise. It must identify the biases associated with claims of truth by those with power over others, and welcome the many voices that are silenced through dominant structures. Truth and reconciliation cannot happen without decolonizing methodologies (Smith, 2012). Decolonized methodologies are more inclusive and open to differences in what constitutes evidence. Recently, the State of California adopted a decolonized approach to youth mental health and innovation, distributing $429 million to hundreds of communities to implement and evaluate either evidence-based practices (EBP) or community-defined evidence-based practices (CDEP). The Salmon et al. chapter in *Stepped Care 2.0: The Power of Conundrums* focused on EBPs. One alternative that we have proposed for years is "practice-based evidence." This opens the door to innovation and experimentation with tracking in real-time outcomes for promising new methods that have not been subject to clinical trials. Practice-based evidence is not sufficient, however, because it is unlikely to invite a wide enough range of ways of knowing. CDEPs, on the other hand, are aimed at doing just that. CDEPs have been defined as:

> a set of practices that communities have used and determined to yield positive results as determined by community consensus over time and which may or may not have been measured empirically but have reached a level of acceptance by the community (Martinez, 2008).

CDEPs recognize the value of accumulated knowledge, expertise, and wisdom of communities throughout history, drawing on prior success and lived experience to inform the design and development of holistic best practices (Martinez et al., 2010). Criteria for CDEPs are clearer, but more loosely defined. For example, it is important to describe clearly the process used to identify a practice as one the community values and endorses, while outcomes are more broadly defined to include anecdotes, and indicators of satisfaction alongside any Eurocentric measures that are valued by the community. In this context, what constitutes evidence is more inclusive and flexible. But more importantly, there is broader accountability to all stakeholders, not just researchers with typical scientific credentials. Rather than competing to determine whose theory or study is more evidence-based, a synergistic approach to evidence allows for the weaving of knowledge, wisdom, values, and uncertainties in a manner that is acceptable to the diverse members that make up a thriving community. This weaving of epistemologies facilitates a process of truth and reconciliation between colonizers and colonized.

Harnessing the Power of the DEI Conundrum Through Awareness

SC2.0 is an attempt to do things differently. However, it still operates within the larger context of healthcare and educational bureaucracies that are rooted in Western ways of thinking that seek to occlude or marginalize Indigenous philosophical worldviews. This is the "default factor" of mental healthcare. Much as the system might pay lip service to the language of diversity, it tends to resist any intervention that might require real change. We raised this conundrum because we believe it is one of the most significant challenges facing any clinician or organization that seeks to place synergy at the heart of their practice.

The conundrum of the "default factor" as it relates to DEI is especially challenging because it affects people on a subconscious level. Many well-meaning practitioners, even those who are vocally supportive of DEI initiatives, do not realize that positivistic and scientistic forms of truth are fundamentally opposed to synergistic and Indigenous healthcare practices. Their unconscious bias toward the Western philosophical framework makes it very difficult to even have a conversation, let alone an academic discussion, about what an alternative way of providing care might look like. As the SC2.0 model expands into new organizations and jurisdictions, it needs to do more than just communicate the advantages of a synergistic approach, as seen through the established metrics of success. It needs to challenge the capitalistic, managerial worldview these metrics come out of and push for a paradigm change.

In Berry's chapter on the conundrum of the default factor, she argues that synergy is the solution we need to the problems the mental healthcare system is plagued by. Synergy could also be the way we get to that solution. Once people experience the power of synergy, it changes them—they do not want to go back to the rigid, hierarchical, dehumanizing approaches to healthcare that are normalized by the Western paradigm. Just as it took centuries for the modern system to be built, it may well take centuries for us to build something new to replace it. If we can get there, it will be because we treat synergy not just as an end goal, but as a means as well.

But how do we bring synergy to the default system? The answer is simple. It is already all around us. While the care system default is scarcity-based, and firmly rooted in capitalist structures, synergy is literally in the air we breathe and the ground we walk on. It is in nature. We can open ourselves to it anytime we want to. Despite many policies and processes aimed at simplifying, reducing, and separating the parts that make up the whole, synergy still emerges. For example, it continues to emerge in group psychotherapy healing when one person's step toward recovery becomes infectious. When one receives a gift in the group, the healing energy expands in a way that all benefit. So, the key to working with the default factor is to seek out synergy that persists and continues to thrive despite forces that work against it. The strand of synergy that embodies awareness reminds us to attend to, draw from, replenish, and sustain the synergy that exists all around us.

Harnessing the Power of the Adaptability/Fidelity Conundrum Through Flexibility

What does SC2.0 look like in practice? This might seem an odd question to ask, given that SC2.0 is a model for care that has already been adopted by several provinces and territories, as well as numerous universities, across Canada and the United States. Still, as Alexia Jaouich, Stacie Carey, Kaitlin Saxton, Joshua Rash, and Jacquie Brown argue, the answer is more complicated than it might seem. Because SC2.0 is meant to be co-designed with local partners and adapted to the needs of each community it is serving, it is supposed to look different in every place it operates. However, in order for it to make sense as a model, it needs to have certain identifiable core components. Otherwise, how will an organization know whether it is implementing SC2.0 with fidelity?

The conundrum Jaouich et al. raise is that of adaptability and fidelity. These two principles are sometimes seen as being in opposition, but they suggest that "when implementing SC2.0, adaptability *is* fidelity" (Jaouich et al., 2023). Jaouich et al. propose that the core components of SC2.0, which are an articulation of its principles and philosophical outlook, are the heart of the model. The question of whether or not SC2.0 is being adapted with fidelity should be answered by reference to the core components, rather than by a checklist of services or modalities being offered.

Flexibility is the strand of synergy that is perhaps most important for this process of assessing fidelity. Administrators who are looking for a simple guide to understanding whether or not they are doing SC2.0 "right" have failed to grasp that successful implementation of SC2.0 means paying attention to how the key parts of the system are grounded in culture-specific identities, processes, and resources that are unique to each distinct community. When SC2.0 works well, access is expanded, and outcomes improve because the community being served has been given space to tailor their care to their needs, and clients from all demographics are empowered to seek out healing. If the principles of social justice, agency, respect, and recovery are being honored, SC2.0 is being implemented with fidelity.

The core components of SC2.0 are guidelines for seeding or tapping into healing synergy. Organizations will attend to the core components that best complement their existing care systems. Adaptations to the implementation and change management work will depend on how synergy already flows in a specific community. Not all core components will be relevant or a priority for those implementing. Just as help seekers might only wish to choose one or two items from the buffet of options on the continuum, it is okay for those implementing a variant of SC2.0 to focus on one or two components. For example, at UC Berkeley, less emphasis has been placed on disrupting the risk paradigm (Core Component 3), given how deeply risk management, credentialing, and reputational management are embedded within the institutional bureaucracy. Synergy does not currently flow well in this area, and as such, the focus will be on addressing conundrums in areas of greater readiness, like the gold standard, DEI, leadership, evidence, intensive care, and stubborn hierarchies.

Harnessing the Power of the Risk Factor Conundrum Through Diplomatic Disruption

According to the standards of practice embraced by most hospitals, student mental health clinics, and licensing boards, licensed mental healthcare providers must abide by strict guidelines. While many of these guidelines were designed to protect clients from the risk of abuse at the hands of powerful professionals, in practice, they tend to serve a different purpose: to protect institutions from lawsuits. Practitioners are encouraged to avoid risk at all costs by adhering to protocols—even if these protocols make it difficult to provide the best quality of care for their clients.

Justin Jacques and Kami Sidman identify the "risk factor" as a major conundrum facing SC2.0. A synergistic healthcare system that empowers clients, respects agency, and maintains that professionals do not have all the wisdom or all the answers is a system that will need to be open to a certain amount of risk. The current risk-averse healthcare structures infantilize clients by stripping them of decision-making power, but are also reluctant to take legal or moral responsibility when abuses do occur. An obsession with minimizing risk warps how we conceive of healing, and ensures that the question "Can I be sued for this?" is at the top of the practitioner's mind whenever they are asked to make a difficult clinical decision. In this context, SC2.0 will always struggle to create a more synergistic way of working.

It is not easy for licensed professionals to accept that they may need to accept a greater degree of risk in their work, and here synergy needs to operate on two different levels. It must be used to help articulate a different way of working, in which clinicians feel empowered to take more risks in supporting their clients, and the strands of diplomatic disruption, contextualization, and awareness need to be put at the center of how risk is negotiated. Rather than imagining that a risk-free way of working exists, organizations must re-imagine risk in a synergistic way by accepting that there will always be risks, and sometimes the only meaningful choice one has as a practitioner is between the risk of a lawsuit and the risk of seeing long-term harm done to a client.

Risk aversion is deeply entrenched in most American healthcare systems. This is not surprising given that policies are typically shaped either by legislative lobbies, or costly litigation. It literally pays to be risk-averse. While Core Component 3 is not a focus of care system building at UC Berkeley, as noted above, diplomatic disruption is still in play. A seasoned diplomat knows when to push, when to pull back, and when to shift allegiances. While the non-academic health services department was not ready to invest in a continuum of care portal for a variety of reasons, including cost and data privacy risks, academic leads in the school of business are willing partners, viewing the model as ripe for evaluation in the context of plans to establish a center of excellence for workplace mental health. The continuum of care portal can be built and tested as part of a research project, but it is not a fully endorsed program for all students. Empathy is woven into the strands of diplomatic disruption, and Peter had to understand the hesitancy to implement the program in the context of the care system from the position of administrators tasked with maintaining the high-status reputation of an elite public institution.

Harnessing the Power of the Gold Standard Conundrum Through Contextualization

The one-at-a-time (OAAT) approach to counseling is one of the hallmarks of SC2.0. Also known as single-session therapy (SST), OAAT treats each counseling session as if it could be the only session, and prioritizes meeting the needs of the client in a single drop-in session. This way of delivering therapy addresses the reality that 70% of clients will only attend a single counseling session, and that repeat weekly ongoing counseling is unlikely. Rather than building a counseling system around the assumption that clients will undergo a lengthy course of treatment, proponents of OAAT argue that we need to be training therapists to provide meaningful, high-quality care for a single, easily accessible visit.

This approach to care has been shown to dramatically decrease wait times and provide more care to more clients. But it has been dogged by misunderstandings: The words "single session" have led some clients and practitioners to believe that *only* one session will be offered. While the phrase "one-at-a-time" is intended to address this misconception, it has faced challenges of its own. Many practitioners and clients remain attached to the idea that "gold standard" care involves regular, ongoing sessions of talk therapy with a licensed professional. Anything less represents inferior care.

In raising the conundrum of the "gold standard," Heather Hair, Arnie Slive, and Monte Bobele engaged in a spirited conversation about the uses and limits of OAAT. All three agreed that OAAT was a marked improvement on the standard way of delivering therapy, but they disagreed on the extent to which OAAT could be used outside of a clinical setting. Hair argued that many people who are not licensed mental health professionals (youth workers, community support providers, activists, and soup kitchen volunteers) essentially do OAATc as part of their work engaging in helpful conversations, but she laments that they are not recognized for the therapeutic help they provide.[1] Expanding OAATc training to these workers could help provide them with tools to do this work more effectively. For Slive and Bobele, who have spent their careers in more clinical contexts, the distinction between licensed and unlicensed professionals still has value. The conundrum of the "gold standard" is not only about how therapy should be delivered, but also about who can provide care.

As an approach to counseling that values flexibility, agency, creativity, and growth orientation, OAAT is one of the most synergistic methods currently available in modern mental health systems. But could it be even more synergistic? The emphasis on licensed professionals is rooted in a way of thinking that assumes one's qualifications to heal can be determined by a university degree and a licensing exam. A truly synergistic approach would accept that everyone holds knowledge and wisdom, and that anyone can be a vessel for healing to flow through. This does not mean no distinctions should be made: Not everyone is called to be a healer, and some people have a greater capacity to heal than others. But a synergistic mindset, rooted in truth and reconciliation, would acknowledge that some people who are called to be healers have not been given the opportunity to attend university or sit a licensing exam. This should not mean they do not have access to tools and training that can make them more effective in their healing work. Add to this the strand of contextualization: Within a broader system of care, with the participation of multiple forms of expertise and unlimited variations of one-at-a-time thinking, there is room for abundant healing practices. Windy Dryden (2019), a student of the Godfather of the prominent Rational Emotive school of cognitive behavioral therapy, argues that single-session care works best if:

1. It is integrated with other services within an agency.
2. Client choice among the services is supported by giving accurate information about the variety of available service options.

Practicing synergy as it relates to OAAT/SST means acknowledging that the needs of clients are various. Expanding the range of options available, and clearly communicating with clients that the path to well-being can take many forms, is essential.

Harnessing the Power of the Stubborn Hierarchies Conundrum Through Creativity

"Nothing about us, without us" is one of the fundamental principles of truth and reconciliation. In the dominant Western paradigm, decision-making is imperialistic: The people at the top of the hierarchy make the big decisions about how their

[1] OAATc is an adaptation of OAAT indicating that non-licensed care providers can be trained to engage helpful one-at-a-time conversations, the "c" representing "helpful conversations" as about to counseling or psychotherapy.

society will move forward, and the people at the bottom of the hierarchy have to find a way to live with it. "Nothing about us, without us" is a call to include the people most affected by a decision in the decision-making process. In the context of SC2.0, this principle is captured in the emphasis on co-design. A central part of the implementation process is the series of workshops and meetings where all stakeholders are brought together to discuss the services and types of care the model will include when implemented.

Marion Cooper raised the conundrum of hierarchies in the co-design process. While co-design is meant to provide an equal space in which every stakeholder speaks with the same authority, whether they are a psychiatrist or a person with living experience, in reality, professional hierarchies (like hierarchies of race, gender, class, ethnicity, and ability) are always present. Some participants will come into the room confident in themselves, their opinions, and the solutions they want to propose. They will speak in accents that are perceived as being authoritative, and will dress in ways that signal their social class. Others might feel out of place or tokenized, or they may feel active hostility for some of their fellow participants, who they might see as being beneficiaries of a broken system. Co-design does not work without honesty and trust, but these are hard to build when everyone understands that outside the co-design space, some people have far more power than others. Cooper argues that successful co-design acknowledges these hierarchies and finds ways to actively counter them by addressing the "elephant in the room" early on, and providing enough time for the important conflicts to come out into the open so they can be dealt with constructively.

Synergy is idealistic, but it is not utopian. It accepts that conflict is part of life. Rather than pretending that power imbalances do not exist, it finds creative and diplomatically disruptive ways to acknowledge the reality of these power imbalances so that they can be at least partially dismantled. Leaders who may find SC2.0's efficiency appealing, but who see co-design as another box to be checked, need to be confronted with and made to understand the central role co-design has to play in the SC2.0 model. Co-design facilitators need to find creative ways to make participants comfortable enough to express their true feelings. Participants need to own their agency and embrace a growth orientation. Synergy is the core of co-design. Without it, co-design will be unable to deliver the lasting change it promises.

There are likely endless creative approaches to address the stubborn hierarchy conundrum. Just as there are more expansive approaches for identifying what constitutes evidence (see the first point in this section), the roles, responsibilities, and activities for supporting mental health can be broadened and more widely distributed. At UC Berkeley, a graduate student on a Fulbright Fellowship (researching the transatlantic slave trade) copied me on an angry email sent to many senior administrators, including the Office of the President for the 10-campus University of California system. Adrian (not his real name) was understandably frustrated by the fact that policies and supports for neurodivergent students like himself were lacking. This was especially concerning for Adrian, given that no community-based specialist services were available in the Bay Area of Northern California. Later, in a separate email to me, he said that he was disappointed by the care he received from our department. I invited him to meet with me on a Zoom conference. When the video image appeared on my screen, I was greeted by a smiling, bright, polite young man who spoke with energy and optimism about his vision for a better care system. I listened to Adrian intently. He had many creative ideas and boundless energy. At our second meeting, I asked if he would be interested in forming a small working group to improve neurodiversity mental healthcare. He was! With living expertise with dyslexia, ADHD, and autism, he brought unique perspectives to our care programming development process. A year later, the university appointed me and one other administrator to lead a Strategic Initiative Committee on Neurodiversity. An idea occurred to me: What if we appointed Adrian as the Committee Chair and served ourselves as "co-mentors" instead of co-chairs? While this disruption was met with some puzzlement, many saw it as a genuine effort to empower students with living expertise to lead the change. As co-mentors, we could offer coaching on navigating the bureaucracy and ensure that Adrian was acquiring leadership skills in an area of great importance to him personally.

Harnessing the Power of the Leadership Conundrum Through Interconnectedness

The hierarchical decision-making model that dominates all modern institutions, in which executive power resides in a handful of positions at the very top of an organization, effectively means that veto power is held by the individuals who are most likely to find a synergistic way of working to be a threat to their authority. Even if the current manager, director, or CEO embraces synergy, if they retire, relocate, or leave the organization, their replacement may decide to undo the work they have done. In fact, it is likely that they will: the cult of leadership in the Western world rewards those who position themselves as agents of change and disruption, regardless of whether the changes they push for are beneficial to the organization itself.

In *Stepped Care 2.0: The Power of Conundrums*, Janis Campbell and Gillian Berry identified the problem of leadership as one of the greatest threats to lasting change. Many of the other conundrums deal with ingrained habits of mind that can, perhaps, evolve over time. But the leadership factor is the one that will determine whether or not an organization has the time and resources to enact change in the first place, and can easily disrupt or disable any long-term change that is underway. This makes it one of the hardest conundrums to work through. How do you apply synergy to a problem—dysfunctional and authoritarian leadership—that is anti-synergistic at its core?

There are no easy answers to this question. But synergy does provide us with some ways of thinking through it. First and foremost, synergy reminds us that individuals working together can wield a lot of power for change. Identifying where the levers of power in an organization are, and how the people with the most access to these levers can be brought on side with the synergistic vision, can help generate useful strategic ideas for exerting pressure on those in power. Second, clinicians who have been trained in synergistic practices may still be able to apply the insights they have gained even if the institution remains hostile. This could be as simple as changing how one approaches each session with a client, or it could involve pushing for a more nuanced and synergistic understanding of DEI work. The potential for synergy is always present. Remembering that every clinician has agency, just as every client has agency, is especially important when working under leadership that is hostile to the ideas and practice of synergy.

A common refrain from some social justice-oriented UC Berkeley staff members was that it was time to break down the hierarchical leadership structures in favor of matrix frameworks. To model such a horizontal shift, the Director (Peter) relinquished his hold of the Student Mental Health "command structure" (McChrystal et al., 2015) in favor of co-directorship. A co-director for innovation, development, and operations was established, making room for a second co-director of wellness and care systems. Secondly, instead of having new staff assigned to managers who had the least number of reports, a new process of assignment was developed to ensure alignment of interests. Interconnectedness was introduced so that staff could report to supervisors from different disciplines who shared similar interests related to innovation and program improvement. As you can see in Fig. 2.2, we are currently moving toward a flatter "team of teams" matrix structure with some staff cross-trained and embedded in adjacent teams. As restructuring is nearing completion, we are entering an alignment phase aimed at preserving the unique identities of distinct teams while eliminating duplication and fostering collaboration, the development of shared policy, and more seamless integrated care for students seeking services.

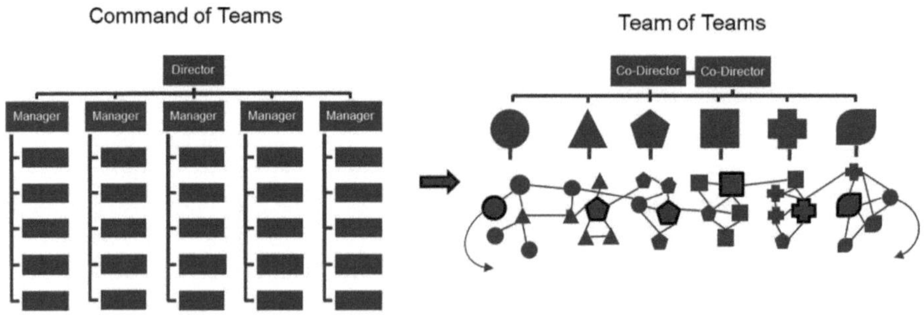

Fig. 2.2 Addressing the leadership conundrum through the interconnectedness of a team-of-teams structure

Harnessing the Power of the Recovery Conundrum Through a Growth Orientation

The word "recovery" is ubiquitous in the world of mental health and substance use services, and while everyone agrees that recovery is a good thing, what is meant by the word can be very different. For some, recovery embraces the harm-reduction approach, which respects individual choice and seeks to maximize well-being as the client understands it. For others, it is synonymous with the abstinence-oriented approach, which removes individual choice and emphasizes the importance of ceasing the use of harmful substances altogether. These two meanings of the word "recovery" have had an impact on other aspects of mental health thinking and practice as well, with the harm-reduction approach to recovery being invoked by clients diagnosed with chronic forms of mental illness who want to emphasize that wellness/illness is a continuum, rather than a binary. But while the mental health sector has moved toward the harm-reduction approach, the way it is understood in the substance-use health sector is more complicated, and most of the public, including political decision-makers, associate the term with abstinence.

As Mary Bartram notes, this creates a conundrum for SC2.0, a model that uses the language of recovery to describe how it approaches mental healthcare. Governments or institutions seeking to implement SC2.0 may be attracted to its embrace of recovery principles because they understand these principles to be abstinence-oriented. SC2.0 could become embroiled in political debates about the use of safe injection sites, or it may risk alienating the substance-use health sector by using the term without an appreciation of the different nuances it carries.

Clarity is important when communicating complex ideas to new audiences, especially when these ideas touch on topics as emotional and polarizing as substance use. But clarity can also be misleading. Imagine, for a moment, the idea that something has 100% clarity. What does that mean? Does it mean 100% certainty? Who finds it clear? How are they certain? Uncertainty and ambiguity leave room for other perspectives. Because SC2.0 is meant to be owned by the communities that adopt it, flexibility is of utmost importance. Synergy respects the basic truth that when we come together in our differences, we can create something greater than each of us as individuals. A synergistic approach to the recovery conundrum acknowledges that an Indigenous community in the Canadian north that has experienced decades of traumatic substance abuse in the wake of colonization might have a different conception of recovery than a clinic working with an economically marginalized and mobile population of unhoused people in a large city. While it is important to strive for as much clarity as possible when communicating with governments and healthcare bureaucracies, it is also important to stress that SC2.0 does not exist to impose its own view of how things should be done in the communities it enters. It exists to empower the members of these communities who are already providing care, and to facilitate greater cooperation between all providers of care. This means that instead of having a single position on recovery, it must acknowledge the diverse views of everyone who is participating in the wellness movement.

So, if full clarity and positionality are unrealistic (and sometimes unproductive), how might synergy help us with the recovery conundrum? The growth orientation strand of synergy is important for recognizing that all perspectives can be accommodated. Sustainable growth does not just mean change, nor does it mean unlimited expansion or dominance. It means adapting for co-existence and complementarity. It nurtures interdependence, respect, and, as much as possible, a place for differences.

Carl Erik Fisher is an addiction physician, bioethicist, writer, and a person in recovery. In addition to his appointment as an Associate Professor at Columbia University, he is a fellow of the American Psychiatric Association. He hosts a podcast called Flourishing After Addiction. Fisher insists that there is a false dichotomy between harm reduction and abstinence. Instead, he argues, addiction is multifaceted and best seen along a continuum. It is "a brain disease, a spiritual malady, the romantic mark of artistic sensibility, a badge of revolution against a sick society, and all of these things at once" (Fisher, 2022). Abstinence is a harm-reduction strategy and exists at one end of the continuum, followed by reduced use, safer use, safer environment, safer supply, and prevent death. With a growth orientation, all points on the continuum have value. In contrast to the more dualistic and judgmental scarcity paradigm (i.e., good/bad, win/lose), Fisher's synergistic perspective expresses infinite possibilities for growth, healing, and recovery.

Harnessing the Power of the Intensive Care Conundrum Through Empathy and Agency

The final conundrum, identified and explored by Andrea Levinson, looks at cases where the client's mental illness temporarily impairs their ability to exercise agency. In the higher steps of the SC2.0 model, where clients may require a stay in hospital or other more intrusive forms of care, the philosophical foundations of SC2.0 are put in creative tension with the realities of persistent, severe, and complex forms of mental illness. Levinson argues that even clients living with serious mental illness can exercise agency with the right supports, and that the SC2.0 approach can provide them with life-changing forms of community and peer support that might not otherwise be available to them.

Care for these clients is not, however, completely in the hands of the mental health system. The ongoing stigma against mental illness means that clients living with mental illness often experience many forms of marginalization that, if addressed, could dramatically improve their quality of life. To improve outcomes for this population, SC2.0 needs to partner with organizations beyond the healthcare system to change the way mental illness is experienced and perceived on a social level. The conundrum of intensive care and serious mental illness points to the fact that mental health is not something that exists in a vacuum. It is shaped by economic pressures and deep cultural prejudices that must be confronted if outcomes are to be improved.

By reminding us of our interconnectedness, and by stressing that agency, strength, and capacity are things everyone possesses in some measure, synergy provides a way forward. But perhaps more important, when working with the most seriously mentally ill, is the strand of empathy. Empathy comes from a place of love, compassion, and grace. When agency is compromised due to grave illness or incapacity, this strand must remain strong. Some aspects of mental health cannot be dealt with in the healthcare system alone and require clinicians to reach out to a wide range of activists and community groups. Over the last two decades, Flexible Assertive Community Treatment (FACT) teams have emerged, which include peers, family members, and other community members in treatment planning and the development of a healing ecosystem. These teams support family and community members in ways that can sustain their involvement in care, preventing burnout by offering helping skills training or respite. A FACT team approach recognizes that sustaining empathy for people who are on the margins of society, especially those who refuse help, is difficult. A synergistic approach to healing communities allows the responsibility for difficult work to be shared. It is easier to love and care for someone who cannot easily reciprocate if the burden is distributed, and a healthy social network ensures reciprocity.

Dutch psychiatrist Rene Keet has implemented, as his default practice, the inclusion of a peer support specialist in psychiatric interviews and consultations. This also seems to us to be a move toward extending the responsibility for care to people who have been there, who can, by their shared experience, offer a different kind of empathy from a skilled psychiatrist who might be more focused on stabilization, medication management, and maintaining safety than on empathy or connection. For the service user, witnessing a peer in a helping role signals a possible pathway for increased agency later in their recovery healing journey.

Synergy does not just need to operate within clinics, hospitals, and mental health centers; it also needs to operate on a community, regional, and national level. Mental healthcare providers need to engage in conversations around housing and minimum wage law, or they might need to work closely with human rights lawyers to achieve meaningful change. Sometimes synergy could mean marching on city hall, or advocating for the rights of refugees. Only by overcoming the artificial silos of our Western cultural mindset can the large-scale power of synergy be let loose.

Conclusion

There is an old Indian parable that illustrates why synergy is so important. A group wearing blindfolds comes across an elephant for the first time, and tries to figure out what sort of creature it is. The first person puts their hands on its trunk, and says it must be a kind of snake. The second puts their hands on its leg, and concludes it is a species of tree. The third puts their hands on its flanks, and insists it is a wall, while the fourth grabs hold of the tail and says it is just a bit of rope. Each person has explored part of the whole, but only together can they understand what the creature is really like.

Our mental healthcare system has, for a long time, privileged one type of knowledge and one way of seeing over all others. So, it should not be surprising that it struggles to comprehend a model like SC2.0, which is predicated on the idea that there are many ways of knowing and many paths to healing. The very things some critics find hardest to understand about SC2.0—the difficulties in measuring its effectiveness, for example, or the lack of a single, quantifiable active ingredient—are the things that make SC2.0 synergistic. This means that SC2.0 will only have a lasting impact on mental healthcare if it can shift the paradigm and provoke a serious conversation about the role of synergy in healing. In responding to the conundrums, we have tried to place them in the context of synergy as a worldview. Synergy requires a paradigm shift, and in responding to a conundrum like the risk factor, or recovery, we must also respond to the philosophical framework these terms come out of—a framework that privileges the sovereign, monadic individual whose identity exists within a series of binaries (well/ill, rational/irrational, man/woman, free/unfree, etc.) over the fluid, interconnected, and indeterminate, and which has its roots in colonialism, white supremacy, ecocide, and genocide. The very things some critics find hardest to understand about SC2.0—the difficulties in measuring its effectiveness, for example, or the lack of a single, quantifiable active ingredient—are the things that make SC2.0 synergistic. This means that SC2.0 will only have a lasting impact on mental healthcare if it is able to shift the paradigm and provoke a serious conversation about the role of synergy in healing.

The conundrums explored in this chapter all arise from a failure to understand or to practice synergy. Organizations that turn to SC2.0 because they want a more efficient way of providing care, or want to reduce their waitlists, or simply need to demonstrate to their stakeholders and governing bodies that they are trying to do something about mental healthcare, need to understand that the model is not a package of features that can be selected or rejected at will. It is a philosophy of well-being rooted in Indigenous ways of knowing and healing practices that have been established for millennia. Only through concrete measures to train staff in synergistic ways of working, like the action steps outlined above, can SC2.0 be implemented successfully.

As mental health service administrators, we often find ourselves caught in mental health crisis narratives. In the context of a scarcity paradigm, the narrative is quite effective in attracting funding that promises to put out the fires. When resources are perceived as scarce, triage is a logical response. But when adopting a synergistic perspective, where healing resources are all around us if we intentionally tap into them, the narrative shifts away from one of crisis to one of opportunity. The work then can focus on improving everyone's access to existing sustainable resources rather than simply investing more on expensive programs that only serve slightly more people with the most urgent needs.

There is no doubt that we have an access problem. There are resources, but we have discounted them for a variety of reasons. This is a consequence of our focus on professionalizing all forms of well-being work while largely ignoring opportunities to address the social determinants of health and well-being. The crisis is less a mental health crisis than a crisis of disconnection from each other, the land, and a sense of purpose, meaning, or grounding in something larger than ourselves. A synergistic perspective can help us expand what we include as healing resources. The SC2.0 principles and framework can help us to tap into synergy by providing an infrastructure that organizes and integrates formal and informal resources, while also attending to policy reform aimed at addressing the social determinants of health.

As we shall see in the next chapter, however, there are many obstacles to building and implementing more synergistic structures in the context of a modern healthcare bureaucracy. Synergy is a countercultural phenomenon, and institutions that are founded on positivist and materialist assumptions do not always have an easy time accommodating changes that are synergistic in nature. The conundrums helped point us toward synergy as a solution, but before we can explore what synergy looks like in practice, we need to go deeper into the philosophical disconnect that arises when trying to bring synergistic thinking to a bureaucratic institution.

References

Cornish, P., & Berry, G. (Eds.). (2023). *Stepped care 2.0: The power of conundrums*. Springer. https://link.springer.com/book/10.1007/978-3-031-45206-2.

Dryden, W. (2019). *Single-session therapy: Distinctive features*. Routledge.

Fisher, C. E. (2022). *The urge: Our history of addiction*. Penguin Press.

Jaouich, A., Carey, S., Saxton, K., Rash, J., & Brown, J. (2023). Adaptability and fidelity: Exploring the relationship between the two when implementing SC2.0. In P. Cornish & G. Berry (Eds.), *Stepped care 2.0: The power of conundrums*. Springer. https://doi.org/10.1007/978-3-031-45206-2_11

Katz, R. (2017). Indigenous healing psychology: Honoring the wisdom of the first peoples. Simon and Schuster.

Martínez, K. (2008). Culturally defined evidence: What is it? And what can it do for Latinas/os? *El Boletin (Newsletter of the National Latina/o Psychological Association)*, Fall/Winter.

Martinez, K., Callejas, L., & Hernandez, M. (2010). Community-Defined Evidence: A Bottom-Up Behavioral Health Approach to Measure What Works in Communities of Color. *Emotional & Behavioral Disorders in Youth, 10*(1), 11-16.

McChrystal, S., Silverman, D., Collins, T., & Fussell, C. (2015). *Team of teams: New rules of engagement for a complex world*. Penguin Random House.

Smith, L. T. (2012). *Decolonizing methodologies: Research and indigenous peoples* (2nd ed.). Zed Books.

Chapter 3
Synergy and Institutions: Why Is Synergy So Hard to Build?

André Forget

Introduction

The concept of synergy is deceptively simple: it is the working together of multiple individual inputs to create an outcome that is more than the sum of its parts. We encounter synergy all the time, and the systems in which we live are all, on a fundamental level, synergistic to one degree or another. Just think of the public transportation network of a large city like New York: Through the collective effort of a relatively small number of maintenance staff, engineers, technicians, drivers, and administrators (each with their own particular role), millions of people travel throughout the city in relative comfort and safety at any time of day. This kind of synergy, rooted in large-scale, centralized coordination, has been and continues to be one of the main drivers of modernity.

The definition of synergy being explored in this book, however, is more particular in that it prioritizes respect for the uniqueness and value of each individual involved in the synergistic process. Emerging from the work of Richard Katz, and drawing on long-established traditions of Indigenous philosophy and psychology, this kind of synergy recognizes the energy that can be unleashed when individuals are treated not as cogs in a machine, but as co-creators and active participants. Stepped Care 2.0 (SC2.0) is inspired by these traditions, and has made synergy a fundamental element of how it conceives of mental healthcare. Without this kind of synergy, the SC2.0 model cannot function. As discussed in the previous chapter, SC2.0 has identified 10 strands that weave together to guide synergistic practice, forming a resilient talking circle (Fig. 3.1). As Katz insists, synergy is fundamentally both a natural and spiritual phenomenon, one that literally grounds us in the land-based ecosystems from which we come and elevates our sense of purpose and

A. Forget (✉)
Sheffield, South Yorkshire, UK

P. Cornish, G. Berry (eds.), *Synergy*,
https://doi.org/10.1007/978-3-031-88493-1_3

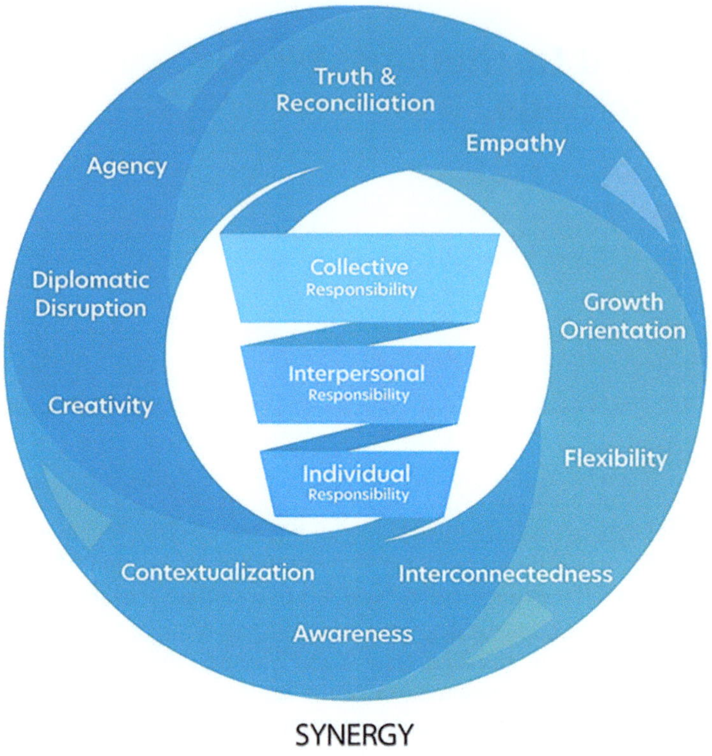

Fig. 3.1 The strands of synergy

meaning to something that exists beyond us. Peter Cornish suggests that the white space within and surrounding this circle can be imagined as containing natural and metaphysical forces beyond our comprehension, unseen, untestable truths, and spiritual healing energies that shape the depth of our souls and very existence.

An organization that operates according to these synergistic principles recognizes the old truth that people are always strongest when working together, but also acknowledges the importance of respecting the individuality and difference that each person brings. Neither collectivist nor individualistic, synergy accepts the tensions, challenges, and risks inherent in living and working with others, and sees them as being a potential source of strength.

When the SC2.0 model is able to channel synergy effectively, it creates a dynamic relationship between leadership, staff, and community—between mental healthcare providers and the communities they serve. This relationship recognizes that many of the binaries that exist within the institutions of mental healthcare and the public (between the sick and the well, between providers and clients, between research experts and research subjects) are unstable, and that true well-being comes when everyone is empowered to give and receive help from those around them. Synergy recognizes that everyone has something to offer. Working synergistically means having the humility to see that the contributions of the C-suite are not fundamentally more important than the contributions of the janitorial staff. The well-being of one relies on the well-being of the other.

This is a stark contrast from the tacit practice of large bureaucracies and institutions in the Western world, in which clear distinctions are made between executives and directors (who are valued for their individual talents, knowledge, and expertise) and the staff (who are expected to adhere to rigid procedural scripts, and who are meant to be easily replaceable). Executives are expected to be extraordinary, and are often celebrated for their counterintuitive views and personal eccentricities—their large salaries are justified by the notion that they have an indispensable vision that will bring success to the organizations they lead. The working people who actually make this vision possible, on the other hand, are generally disciplined if they step outside their carefully bound roles. It is assumed that they serve the organization best by following the instructions sent down from the executive, and should they feel the instructions are nonsensical, or are creating a hazard to health, safety, and even profitability, they will likely be ignored, silenced, or fired. There is a philosophy of synergy in Western institutions, but it is the synergy of the barracks and the battlefield, not the synergy of the Fijian fish drive or the Amish barn-raising.

This distinction is of central importance to the arguments being made in this book: Since it was first coined by the American architect Buckminster Fuller, "synergy" has become a standard buzzword among corporate elites, who have often used it as a pretext for mergers, streamlining, and outsourcing. The more specific sense in which we are using it goes back much further than Fuller, to the many Indigenous civilizations around the globe who understand it as a bedrock principle of social life. This approach to synergy adds an essential dimension—which, for lack of a better word, might be called the spiritual.

The social sciences do not have an easy time with the word "spiritual," rooted as they are in Enlightenment conceptions of reason, materialism, and objectivity. Spirituality in the West has long occupied a fraught place between established religion (specifically the organized, global religions) and skeptical secularism. For that reason, researchers have often preferred to bracket it out. But in this context, it can be taken to mean the quality of simultaneous uniqueness and interconnectedness that each living and non-living thing brings to every process in which they participate. Synergy that ignores spirituality has a tendency to become technocratic and authoritarian, because it fails to see the uniqueness of each component, instead imagining that the components can be replaced or sacrificed without doing fundamental damage to the whole. Synergy that embraces spirituality has the potential to be less hierarchical, more dynamic, and, in the long run, more effective. It is also, in the context of modern institutions, much harder to achieve, which is one of the reasons that this chapter is dedicated to the problem of how to build synergy in an institutional context. If synergy makes so much intuitive sense, why is not it more widely practiced? Perhaps it is because the corporate view of synergy is actively hostile to the older and more authentic synergy Katz saw at work in Indigenous communities.

We see this clearly in the example of the New York subway system. While the subway is a wonder of modern engineering and an embodiment of the modern approach to rational corporate governance, it is also aging, underfunded, and not always safe for passengers. Why is this? In part, it is due to the reigning neoliberal ideology of our era—an ideology that values profitable assets over public goods. But it is also due to the relationship between the Metropolitan Transportation Authority (the public benefit corporation that operates the subway), its employees, and the people it serves. The MTA operates at arms-length from the government, and while it provides a service that millions of New Yorkers rely on, its passengers have little sense of ownership over it. The relationship is, in fact, more often an oppositional one: Transit authorities are constantly on the lookout for antisocial behavior and fare evasion, transportation staff do stressful work and have to interact with a sometimes-hostile public, and passengers are exhausted and annoyed by unreliable timetables and poorly maintained infrastructure. More affluent New Yorkers rarely use the service at all, preferring cars or helicopters, which further reinforces the city's class differences. While the subway is the quintessence of a public service—something that exists to benefit everyone involved—the hierarchical structure of its organization means that people using and operating the service do not feel much of a sense of ownership over it, and the people in charge are often removed from the day-to-day operations. These different groups see themselves as being in opposition to each other, rather than as being collaborators together. And this is by far the norm in modern institutions. From politics and policing to the arts and the food services industry, institutions are all too often run in such a way as to foster a sense of competition and distrust. No one feels particularly empowered (even among those who have power), and everyone feels that their particular challenges are being overlooked.

What would a more synergistic approach look like? It would need to start by focusing on what the institutions actually exist to do (providing freedom and ease of movement for the residents of a city, in the case of the MTA) rather than on how much profit can be made from doing it. It would also need to take into account the experiences of both the service providers and service users. Where are the points of tension? Can these tensions be resolved through devolution of authority, better funding, or better technology? When a problem emerges—threatening or antisocial behavior on trains, for example—a synergistic way of working would start by acknowledging that the individuals responsible for this behavior are part of the community, too, and would emphasize truth and reconciliation over carceral approaches. Finally, communities would need to see things like the MTA as belonging to them, as being part of their collective rights and responsibilities as citizens of the city.

This would require a major shift in thinking on the part of institutional leaders, and it is far from clear that building momentum for this shift is possible. The purpose of this chapter is to explore the pitfalls of building spiritually inflected synergy in the context of modern institutions and bureaucracies, particularly those that deliver mental health care, so the dimensions of the problem can be seen more clearly. By engaging with the experiences of three mental health professionals, each from a different background and at a different point in their careers, each of whom has an interest in synergistic approaches to healing, we will see the complex ways in which Western institutional thinking, shaped as it is by colonialism, racism, and patriarchy, creates significant challenges for those seeking to work synergistically. We will also explore ways in which these challenges have, in particular times and particular places, been overcome.

Revolutionary Potential and Institutional Backlash

Dr. Genée Jackson is a recently retired senior staff psychologist from the University Health Services of the University of California, Berkeley, who runs a private practice with offices in Oakland and San Francisco. When I first met her in early 2024, I quickly realized I was in the presence of someone who could speak with authority—not the authority of credentials (though of course she has those as well) but the rarer and more genuine authority that comes from a lifetime of experience. During our first conversation, she effortlessly situated my questions about synergy within a larger context of a struggle for social justice (a struggle she has spent her life serving), and while her breadth of knowledge was remarkable, she spoke with a down-to-earth pragmatism. She was generous with her time and insights, and open to the ideas of others, but always willing to push back on points she felt were insufficiently nuanced or analyses that she believed to be insufficiently coherent. She was at ease talking about complex ideas, but her preferred way of making a point was through the telling of a story—as if she was trying to communicate that even the most profound and abstract ideas have their basis in ordinary struggles for dignity, justice, and recognition.

These are struggles Jackson knows a lot about, for she has had a long and fascinating career. Growing up in New York City in the 1960s, she experienced first-hand the White hostility that came from being part of the first group of Black children to integrate public schools. Later, while an undergraduate student at a university in Middletown, Connecticut, she volunteered in the Big Sisters, and participated in the revolutionary community activism of the Black Panther's Free Breakfast Program for children. After moving from Connecticut to the San Francisco Bay Area in 1976, she was thrown into the exhilarating experiment in revolutionary community activism that was still in full swing. The Bay Area in those years was a petri dish of radical experiments in community organization and political activism; for Jackson, it was a time of grassroots organizing, consciousness-raising, advocacy, and political work aimed at undoing centuries of oppression. The Pan-African People's Organization was giving Black Americans new language and strategies for talking about identity, solidarity, and liberation, and the recently formed Association of Black Psychologists was raising profound questions about the Eurocentricity of the establishment approach to addressing the mental health of Black individuals, families, and communities. The community Jackson found there was not only a school for learning about the world, but for learning how to change it. These organizations each provided environments that personified synergy in action.

When I talk to Jackson over a video call, she stresses how important this formative experience in the Bay Area was. "I think it was the place, the time in history, the people...all these things converged in a way that I don't think could have happened anywhere else," she says.

Working as a graphic designer, photographer, and educational media specialist, Jackson had a front-row view of how a well-planned comprehensive Community Mental Health Consortium could make meaningful changes across multiethnic and multiracial communities in San Francisco. "I worked among a group of Black psychologists who understood how to blend research-based clinical treatment and systems theory in conjunction with basic practices in applied community psychology," she recalls. The Westside Community Mental Health Consortium, consisting of 18 agencies in San Francisco, was part of a nationally funded initiative of the Kennedy administration to create comprehensive mental health care centers across the United States. These centers were charged with addressing all levels of care, from prevention programs to acute inpatient care, and the Westside Consortium was seen as a premier example for the entire country in showcasing best practices for providing culturally relevant services across several multiethnic communities. "We met monthly with a Community Advisory Board made up of members from each community to discuss ways to address needs that they identified," Jackson says. "We also collaborated with church pastors, training them to more easily identify those in need of referral to our programs. Westside very effectively blended the use of a systemic lens with classic principles of community psychology and a social justice philosophy to foster a successful model of co-created and collaborative care."

Jackson continued, "During my time there, I was directly involved with several very high-profile events. Westside had a front-seat role in winning the landmark Larry P. case that changed educational testing practices across the state of California. They also conducted the only in-depth investigation into what really happened to members of Jim Jones' People's Temple in Jonestown, Guyana. Those families, and their survivors, were all part of a community that fell under Westside's purview. In both cases, they took on a strong advocacy role and held a clear sense of responsibility for how to systemically address the multiple needs of the people within the diverse communities they served."

Seeing these psychologists in action at Westside was Jackson's first glimpse of understanding how psychologists could translate and apply their training in a culturally sensitive and skilful manner. They collaborated with multiple communities, helping them develop resources for their own people, by their own people; an extraordinary example of what can be accomplished when people feel empowered to take charge of their own lives together. It is a perfect example of synergy in action.

Jackson gained valuable experience in her work as a community organizer, but it was her work as a volunteer at Youth Advocates, a private non-profit agency for teen runaways, that set her down the path toward becoming a licensed clinical psychologist, a path that would lead her to grapple with the question of how that same synergy can be applied in individual and institutional contexts.

"My volunteer work at Youth Advocates led to a job where I was part of family reunification sessions that were often very intense family interventions. I felt like I was doing work that was really hard, nuanced, and complicated," Jackson recalls. "And so it made me decide that if I wanted to continue doing this type of work, I needed to get trained." This decision proved to be a fateful one. Although Jackson grew up in New York City, her family originally hails from Oklahoma. Her mother had to leave to pursue a degree at Columbia University in New York City, because it was easier for Jackson's mother to pursue her education at a prestigious Ivy League institution than it was to attend Oklahoma State University, which at that time refused to admit Black students and, instead, paid institutions outside of the state to take them. By the time Jackson applied in the mid-1980s, it no longer had an officially racist admissions policy, and they had stopped a more recent practice of roping off the classrooms to separate Black from White students. But that did not mean the university, and the city of Stillwater in which its main campus was located, had ceased to be a deeply racist environment. The hostility and violence she experienced there, both on and off the campus, was harrowing, and even though the classroom was not as physically dangerous as the streets turned out to be for a Black graduate student, that did not mean there were not enormous obstacles to overcome.

"My time in training was extremely difficult, due to the racially charged atmosphere on the campus and in the city," she says. "Juxtapose this with the content I was learning, which turned out to be very focused on individual therapy, using European models based on European populations." Because of this, she pursued a second master's degree specifically to gain proficiency in the use of systems theory as it applied to work with individuals, families, and communities. She remembers one class in which the Larry P. v. Riles legal case was being discussed—a case that revolved around the use of IQ tests to funnel Black students into special education classes. Jackson had personal knowledge of the case because she worked with the psychologists who were consulting directly with the Larry P. legal team. She was shocked by how naïve the class seemed to be about the process that had led to this landmark legal decision. "I said, this didn't happen because some kind person in a classroom or courtroom said, 'Hey, maybe we should help these kids!'" she recalls. "I was trying to explain that this came with a considerable fight and was initiated by parents and professionals in the San Francisco Bay Area Black community. This was not part of the status quo system. That professor later pulled me aside to say, 'We don't do that here. We don't speak up like that in class.' He really meant it to be helpful, advising me to be cautious with this in my other classes. Their idea of an education was that you sit, you listen, and don't raise issues or ask questions." This experience confirmed what Jackson had already suspected entering the program.

"If I'm going to work as a psychologist, I can't do it unless what I'm doing involves changing the profession. That's how I felt. Unless this profession can broaden how we as professionals define it and train psychologists to be critical thinkers, capable of advocacy and systemic change, I can't subscribe to it."

Jackson got a chance to put her dedication to change into action when she finally, and with some relief, left Oklahoma for good. Not long after, she took up a position in Alaska, where she founded the Multicultural Mental Health Summer Institute (MMHSI) at the University of Alaska for graduate students and interested community members who wanted to improve their multicultural competencies. By this time, in the 1990s, the importance of multicultural training was starting to gain traction in the United States. The American Psychological Association instituted a new requirement for psychology licensure—a course containing content related to "multicultural psychology." The MMHSI was intended to fulfill that requirement. "It was for psychology graduate students and others preparing for their exams who needed to fill a new and badly needed requirement for licensure, as well as anyone else who was interested—especially those who worked with the local tribal groups—and felt like they needed more skills and more awareness and sensitivity in their work across cultures," Jackson explains.

"Once advertised, it was a shock to the entire psychology department to see the large numbers of interested participants from the Anchorage community who worked in schools, public agencies, and mental health facilities. All were seeking tools for understanding their own positionality vis-à-vis America's long history of racist discrimination. We could not accommodate the numbers! A waiting list had to be created for the following summer.

"One long-standing tenured colleague pulled me aside and admitted that he had no understanding of what the Institute was all about but was very surprised to see there was such a demand for this new information. With genuine surprise, he said I guess there's really something here to know about!"

The material was approached intensively. There was a lot of reading, but there was also a big experiential component where they got to explore what it all meant to them. The responses she received from participants were powerful, and she recalls some of them coming away from sessions having had profound emotional experiences. "They would be in tears. Sobbing and grateful. 'Thank you, this has changed my life'—the kinds of responses that told me that, okay, this is a meaningful way of engaging. Of course, they chose to be there. They wanted skills, they wanted that deeper understanding, and having 'small group' facilitated experiences and facilitated discussions where you were able to work with their issues within the group (and privately when necessary) made a huge difference. But it was an extremely powerful healing experience." It had a powerful influence on the rest of her career, many years of which were spent as a diversity consultant in the government, non-profit, and corporate worlds.

Though the context of Jackson's work in Alaska was very different from her experience more than two decades earlier in San Francisco, both were examples of synergy. Community and academic institutions played a role in the process in both cases, insofar as they created a space where people could work through extremely difficult questions related to identity, heritage, and justice. But when I ask Jackson how she sees the relationship between synergy and institutions, she takes a thoughtful pause before explaining that, in her experience, one of the challenges of facilitating deep, transformative, synergistic change in the institutional context often comes down to how much time and skill it takes. "What allowed for the success of the Institute in Alaska was the institution's openness to new ideas and willingness to offer resources to make it happen. Their bureaucratic systems were less complicated, and I was given almost complete freedom to create the program and curriculum."

Following Alaska, Jackson had a stint as an acting director of a PhD program in Culture and Human Behavior at the California School of Professional Psychology

(CSPP). It was the first doctoral training program of its kind, and it prepared students from the United States and other countries to apply psychological theory and skills with a cultural lens to institutions across industries of business, education, government, and private non-profit institutions. At the time, CSPP was willing to offer its support and allow space for senior faculty members to develop a novel high-level approach to training psychologists to lead with a cultural focus while working systematically in institutional settings.

"Unfortunately, a lot of inaccurate assumptions are made by larger bureaucracies," she says. "They think that anybody can do it—it's like, oh yeah, just get people together and let them talk, and we can just get [any employee who is a person of color] to facilitate, because he is Black, or she is Latina, they can run that group. This, to me, is a very disrespectful and ignorant approach. And it's designed to fail, because that person—first of all, he or she is an employee, right? So, the power differential means it's automatically not a safe space. Secondly, they likely don't have the training to know how to facilitate deep material like that. Things can easily blow up in their face, and the group will wind up with difficulties that can have longstanding consequences. I've seen really messy things happen in larger bureaucracies that try to get around approaching it in a correct way, using expertise. They're often unwilling to spend money on trained consultants and don't understand the complexities of the issues."

"When diversity training is done well," Jackson says, "it's because a space is created in which synergy can operate, where the barriers that people have built up over time to protect themselves can be broken down, and they can freely experience the truth of another person's experience, or express their own, without fear of judgment or retribution." This is extremely difficult, especially for participants from privileged groups who may not want to think about how brutal the systems they operate within can be. It requires willingness, patience, and a facilitator who is coming from the outside and is not enmeshed in the corporate or office politics of the organization. Sometimes, in Jackson's experience, institutions have been willing to put in the hard work to really evaluate where their work culture needs to change.

In one case, when Jackson was called to consult with a major museum, she was pleasantly surprised to discover they had laid out an intensive five-year collaboratively designed training plan to be delivered to all employees. The training was customized to the workplace issues and population, was conducted by outside experts, and included an evaluation of their initial status and outcomes. It even had clinicians available throughout the training day for anyone needing support. "It was impressive!" Jackson says. "A great example of an institution taking the task of changing their workplace culture seriously. They were willing to create the needed space and to make structural changes in order to achieve a larger goal."

Sadly, the heart-based work done by expert consultants over earlier decades eventually morphed into corporate-led efforts to turn it into an in-house function referred to as diversity, equity, and inclusion (DEI). They established political, heavily monitored, static positions inside organisations, typically situated in areas without power or authority where all the important elements of the earlier change work were lost, either by design or by default. In comparison, DEI is simply a box-checking exercise, a corporate initiative designed not to change the power dynamics of the organization, raise awareness, or improve the quality of interpersonal interactions, but to ensure that they stay more or less the same.

At one point, Jackson ended up facilitating a diversity seminar for a police department. She recalls the officers sitting in their rows of chairs, arms folded, utterly uninterested in what she had to bring. For Jackson, this typified the shortcomings of the bureaucratic or institutional approach to DEI. By contrast, on another occasion, she was called in by the Oakland Police Department.

"It was collaboratively decided that I make myself available to any officers who chose to meet with me, creating a safe space that allowed for honest exploration focused only on their content and opening a door for a respectful exchange on other topics. What happened? A young officer sought help with managing his stress. The conversation eventually shifted to the department's new rule that everyone would wear a bodycam as a way of monitoring and reducing incidents and accusations of racially motivated police brutality. The White officer was able to safely explore his thoughts and feelings about the internal and external workplace issues pertaining to race."

Real, lasting change to people's outlooks and attitudes cannot be imposed by corporate or government fiat. People need to be engaged, and there needs to be a willingness to participate in the process. This is one of the fundamental challenges of trying to build synergy within institutions, which are so fond of uniformity and predictable results—synergy itself depends on the uniqueness of every person involved, and this makes it difficult to boil it down into a formula that can be bottled and decanted in any context.

Thinking back to her early years in San Francisco, Jackson stresses how specific the circumstances that led to that extraordinary efflorescence of activism, intellect, and solidarity really were. Something special had happened there, something that probably could not have happened exactly the same way anywhere else. She becomes almost wistful remembering the spirit of the counterculture and social justice movements of the 1960s, how passionate her community was, how hungry for change.

"What I miss the most about those years (and people will say this if they're around my age) is the fact that many people identified as social change agents, and they really did lead with their hearts. And when you're working for a larger cause of any kind—it could be the climate, it could be anything—but when you are working on something bigger than you that feels more important than anything else, and you're coming together with other people who have that same passion, it's a heart connection. And as I see it, except for a small minority, that's no longer the case."

The Bay Area community Jackson participated in did not arise from a "DEI" program or a business plan. It emerged from the collective energies and organizational abilities of hundreds if not thousands of individuals. The institutional response to this kind of spontaneous, bottom-up movement is often about either shutting it down, or controlling and replicating it, finding ways to remove the particular human component to make it more easily portable. A synergistic understanding, however, grasps that the movement for racial justice in the Bay Area was fundamentally about the people who are working together there. The goal should not be to copy it or revive it, but to tap into the same energies those activists were tapping into 50 years ago.

Jackson ended the conversation on a positive note. In the early 1990s, when she was an intern at UC Berkeley, she had felt strongly that it was important for her to connect with Black student organizations on campus to better understand and address their needs. But her managers had been largely unwilling to support her in this work. "At that time, it was not seen as an appropriate way to provide psychological services—the service was not structured in a way that allowed counselors to function on the campus outside of the building, outside of one-on-one interactions, nor did they even sanction a focus on the needs of student groups based on their racial or ethnic background." Though she continued to work at Berkeley for several years after her internship was completed, and continued to attempt to bring attention to the particular needs of students of color, it became clear that the workplace culture was not a good fit. And the challenge of developing a multicultural graduate training program was calling her to Alaska.

Thirty years later, however, she returned to UC Berkeley and discovered a very different attitude and philosophy at their health service. New leadership was promoting meaningful discussions about the importance of providing culturally sensitive programs to a diverse student population. In this new context, Jackson was able to propose and co-create an African American Mental Health team alongside three other African American staff persons at the counseling center. "This was during the time George Floyd had been murdered by the police," Jackson recalls. "Our team moved quickly to become immediately responsive to the Black student community's needs in light of multiple political and racial incidents that occurred on the campus and nationally. The programs were customized to student concerns and co-created with university staff and students." This team worked closely with everyone from the student program administrators to the athletics department, and ended up being awarded the Louise Patterson Award from the African American Student Development Office.

"In a very short period of time, with resources and support, this team was extremely successful at creating a strong presence that positively impacted the campus community," Jackson says. "Yes, we did face some internal systemic pushback, but this was a huge contrast to my work as an intern at the same institution, in the same health service, in the same building, and same offices, thirty years earlier. It shows that with intention, institutions *can* allow space for flexibility in responding to unique needs and providing relevant services. But sadly, it took several decades and a change in leadership to get there."

Understanding the Importance of the Spiritual Dimension

Dr. Raghav Suri has, since 2017, been a behavioral health provider at the Health Services Center at the University of California, Berkeley. But before he became a clinical psychologist, he had trained as a homeopathic physician in India. This background has given Suri a different outlook on health and wellness from that of his colleagues who followed the typical Western academic career path, and it is part of what fuels his interest in synergy. As he tells me early on in our conversation, he believes synergy needs to be understood in a larger scientific and philosophical context, and feels that the siloization of Western thought has made it hard for those operating from a purely Western paradigm to do that.

"Western institutions don't know what synergy is," he explains gently. "Synergy is the golden thread—it's rooted in spirituality and understanding of the butterfly effect. It's about how different elements, different parts of the system, are connected and have this much greater connection with the universe. Everything is connected to everything else in the universe, whether it's nature or even just in the fact that we are all connected through energy, and that has a ripple effect. When it comes to these more abstract or spiritual concepts, I think Western institutions have brainwashed us so much that we are almost afraid to even acknowledge the possibility of these things—but then when it comes in the form of quantum physics or other acceptable scientific forms to our current society, we start to embrace them pretty well. To me, the abstract concept of synergy is about the connectivity of one another and how that connectivity either aligns or misaligns."

For Suri, this sense of an interconnected world—a world in which it is nearly impossible, on the level of physics, to talk about an element existing independent of the elements around it—is the essence of synergy. But he believes it has a more pointed, political meaning as well. "Synergy," he says, "represents that lost element of our current society, our institutions, our education system, our way of looking at our overreliance on the tangible thing. If we can see it and we can understand it in a physical format, we will believe it." Suri thinks that the West's overreliance on a crudely materialistic understanding of what constitutes knowledge has rendered thinkers who operate only within a Western paradigm incapable of seeing aspects of life that cannot be neatly captured in data. "There's a difference between something that's disproven versus something that's unproven—but in our current scientific world or our worldviews, if we don't have proof of something, that means it's disproven. And that to me is hugely problematic." This is compounded by the fact that while we often talk about "science" as if it is a single body of knowledge, the discussions being had and the assumptions being made in the realm of, say, theoretical physics are quite different from those in biology. Appeals to science do not always take into account the diversity of thought that exists even within the Western academic traditions of scientific research.

Having grown up in India, Suri is acutely aware that the Western tradition of thought is simply one among many. He argues that its pretension to universalism masks a particular, culturally rooted understanding of the world that has gone global not because it is truer, but because it was exported through a violent process of colonialism. When European powers invaded the Americas, Africa, Asia, and Oceania, they brought with them an understanding of the world in which science and spirituality were distinct realms. Though the Christianity forced on the populations of colonized nations often went hand-in-hand with scientific research, Europeans had a concept of "the secular" that arose from the particular way in which church (the sole recognized spiritual authority) and state had developed in Western Europe following the disintegration of the Roman Empire. By the eighteenth century, the natural sciences in Europe were seen as being a body of knowledge distinct (and often in tension with) spirituality, and this way of conceiving of the world was imposed on cultures like those in India with far older and more integrated understandings of scientific and spiritual truth. The result was not only a concerted attempt to erase and destroy vibrant and ancient spiritual and philosophical traditions; it also means that the modern academic world, which is still dominated by Western institutions, has tended to treat spirituality as a private matter of individual choice, rather than a central aspect of human experience.

That the world of mental health care, which is ostensibly dedicated to dealing with personal experience, does not have a clear sense of where the spiritual fits into things is particularly strange. "In most of our mental health training right now, we don't talk about spirituality," Suri notes. "We have physical, mental, and social health, but we leave out spirituality. Now, the problem with that is that when we are working clinically, and someone brings in their religious or spiritual perspectives, we always refer to the studies that have shown that our beliefs are extremely important in terms of our physical and mental health—but we failed to implement it actually into our intake. A lot of times, even in our initial understanding, we don't integrate it as part of who a person is, and understanding that there is a spectrum of spirituality. So to me, our mental health cannot have that synergy when we are leaving out a portion of human existence."

The problem becomes particularly acute when dealing with deep-rooted issues around identity and purpose, given that most forms of spirituality acknowledge that there is purpose and meaning in human life. "One of the things I love about Hinduism (which is not the correct word; it's a colonial term) is you can be an atheist and still follow the principles," Suri explains. "It's more about the way of life, the way of being, and that very much has roots in defining the purpose of being. And all of those elements, or the Japanese principles of Ikigai, are very similar to me. But this is a conversation I can't even have a lot of times on the Western side of things because I would have to explain how any of this relates to mental health." For Suri, however, the relationship is obvious: Mental health is deeply tied to fundamental questions about what it means to be human. Whether a client is dealing with depression, insomnia, or grief, the issue is almost always related to trauma, intergenerational trauma, or the changes and vicissitudes of life. It is, in short, "all tied to how we navigate the world and to navigate the world, how we understand the world."

This inability on the part of Western institutions and systems to seriously grapple with the relationship between mental health and spirituality is one sign of how deeply un-synergistic the Western paradigm of thought can be. It is also something that makes creating synergy extremely difficult in the context of such institutions. Because spiritual questions that tap into the very core of our identities and personhoods need to be left out of the equation, the ability of people to come together in meaningful ways is significantly hampered. While it might mean less immediate conflict (people with strongly held but divergent spiritual beliefs can work together without trying to impose their views on each other), the price for this surface-level peace is not negligible. "We are hiding parts of ourselves that are not socially acceptable," Suri says, "and then we are also siloing elements and people, and we are moving towards homogeneity, and we are losing ourselves and parts of the society and the system in that process."

When I ask Suri how he thinks synergy can be built in the context of colonial Western institutions, he suggests that focusing too much on the institutions themselves is a mistake. "I don't think we can change the institutions," he says, "because institutions are, from my perspective, an effect, not a cause. Institutions are a reflection of our society. We can have some broad public policies and those things, but we are talking about beliefs here with synergy. And you cannot dictate beliefs unless you go the colonialism route and have a hundred-year plan. But to me, we need to start with changing the people and the culture." While changing the culture might seem a daunting task, he points out that psychotherapists are in a uniquely powerful position to do so. Therapy is, among other things, about education, and when a therapist sees a client, they are not just helping them work through their personal problems, they are also providing that client with a language for thinking about their lives. In a synergistic way, therapists can spread healing through the culture, helping clients see things in a new way and empowering them to share that new way of seeing with their own social circles.

"A psychologist sees, what, a hundred, two hundred people? And those people have a ripple effect on other people in their lives. When we can have those spiritual conversations and normalize them for someone, psychologists are extremely influential." This, he believes, is far more effective than trying to introduce "policies and protocols and systems for the sake of including what seems like a 'diversity chapter' that a lot of institutions have done, or the environmental impact reports that for most companies live on the public relations side of things. So to me, that part of institutional change is meaningless, but the educational part is impactful and meaningful."

Teamwork, Synergy, and Basketball

Kareem South is an athlete, a coach, and an early-career mental health worker. Though still in his mid-twenties, he seems to have already lived several lives—as a college basketball player, as an MA student in public health at the University of California, Berkeley, as a member of the Edmonton Stingers, and finally, for a period of time, as an engagement and social impact associate at Stepped Care Solutions in Toronto. But when I ask him to tell me about his life, it is clear that from his own perspective, these different facets of his life have all been part of a single mission, one nourished at a deep level by his passion for healing the mind and the body, and his abiding Christian faith. Perhaps unsurprisingly, when I ask him about the moments when he is seen synergy functioning in his own life, South turns to his experience with basketball.

"There are moments of harmony and just joy that came from reaching a goal together as a team winning championships and realizing what it actually means to succeed at the professional level," he recalls. "But the moments of synergy take place before you actually raise up the trophy in the banners. It's in the fellowship, the connection—and despite what we did on the basketball court, it was how we bonded and connected as individuals outside of practice, outside of games. There was just a real understanding that we were in this together. So, for me, seeing instances where a game like basketball can translate into this environment where healing and empowerment and agency can be fostered, I thought to myself, well, how can we capture this and bring it outside of the game and into other areas?"

South had spent 2 years playing professional basketball in Edmonton, Canada, and he was eager to take what he had learned back to his community in Toronto. But when he returned to the courts he had grown up on, he found that something different was happening there—something that reflected broader challenges within the community. "The way that people would engage with each other was more combative, and it was almost like their livelihoods were at stake if they didn't win a basketball game," he says. "And I was like, this isn't how we played the game! And that's my opinion, but I realized that what I was seeing manifest on the basketball court was an indication of where people's hearts were in their own personal lives."

In addition to being a game, basketball was also a path for advancement and progress in a world that did not offer many avenues to success for young people in South's community. "People are looking for a way out of their situation," South said. "They want more—maybe they want material fame or wealth, or they want to be seen and heard. They want to establish their autonomy and agency as a recognized member of society. And the messages that come from the mainstream, most often they come from celebrities or people in high positions like politicians or authorities. But it is a message that doesn't speak to a person's truest identity or purpose or meaning."

For South, the analogy between basketball and mental healthcare spoke to a deeper set of problems within contemporary Western society more generally. "There is a right way to play basketball, a right way to win," he says. "But the measures of success are often wrong." Rather than seeing good basketball as being about fitness, teamwork, and the beauty of a well-played game, financial incentives and the desire for personal fame had created a situation where the game itself was less important than the money that could be made from basketball as an industry—a team that plays joyless, competent basketball and suffers from regular injuries but which still fills stadiums and racks up endorsements is just as "successful" as a team that plays beautifully and works well together. South believes a similar form of confusion about purpose afflicts the mental healthcare system.

"The challenge in the mental health system is an incongruity between what we believe will affect change, and what is happening. The system is not delivering what it says it will," he says. "Healing is a process that involves identifying that there is or has been a wound, and knowing how to heal comes next. It requires knowledge of different kinds, but especially experiential knowledge." Unfortunately, the system he was trained to operate in as a student of public health is one where healing is not the only goal, or perhaps even the main one. The commercialization of mental healthcare that developed alongside the clinical practice of mental healthcare in the

West ensures that the same financial incentives that distort professional basketball have distorted the system of healing. "There has been a mass deception regarding what humans believe is important. We have been distracted for a long time, and we're not focusing on things that make us feel whole and sane. This impedes the healing process, and people are suffering because of it."

As an early-career professional, South is sensitive to the challenges involved in addressing this tendency within systems to shift emphasis away from their stated purpose for being and toward institutional survival and growth. This is not just a result of the iron law of institutions, but also a product of a larger culture that has been set up to prioritize individual excellence over the success of the group. "In sports or in business, climbing up the ladder means getting the accolades, getting the degrees, getting the accreditations, getting the sense of acclaim," he says. This kind of success does not always mean doing what is right for the organization as a whole, let alone ensuring that the purpose of the organization is being fulfilled. Rather, it involves a ruthless focus on what is best for oneself, and finding ways to bend the institution to one's will. This often leads to the technocratic synergy discussed at the beginning of the chapter, where some individuals have the authority to command, and the majority must fall in line. This is a perversion of truly synergistic teamwork, because it fails to account for the gifts and talents that less assertive and less ambitious members bring to the process. It also creates fertile ground for the perversion of the mission itself. A mental health professional who is primarily interested in spreading their own theories and burnishing their own reputation may initially be motivated by a genuine desire to help people. But if they turn out to be wrong, and their ideas do not move the needle in terms of creating greater well-being, are they likely to acknowledge that they were wrong, and accept the public embarrassment and diminished career prospects? Too often, they will simply double down on their theories and insist that it is the data that is at fault, because this is what the entire incentive structure of the system has trained them to do.

One antidote to this is a measure of humility. A particularly difficult lesson South learned during his career in collegiate and professional sports was the importance of putting the good of the team over his own personal desire for more time in the spotlight. "Playing on a championship team with really good players, I realized that for the sake of our team's success, I had to take a position that required me to play fewer minutes," he says. "And even in the limited minutes that I had, or the limited role that I had, I knew that I had to maximize all that I could bring to the team even if I wasn't playing, even if I was injured." This is very hard to do, especially if the coach or the team leader does not properly acknowledge the sacrifice that is being made for the greater good. In a context where competitiveness and individual performance are rewarded, and collaboration and a willingness to step back so others can share their opinions and contribute to the vision is punished, synergy requires a complete change in the workplace culture.

Regardless of the difficulties, South believes the synergy of the basketball team provides a lens for thinking about how synergy can be used for diplomatic disruption within institutions. Since leaving professional basketball, he has spent time volunteering as a coach. This experience has taught him the importance of seeing the whole, and trying to foster an environment where the emphasis is on bringing people into synergistic relationships where each individual's role is recognized and valued. "There've been countless stories of friends, mentees, players that I've coached that have essentially slipped through the cracks of our system, for many reasons," he says. "But I think part of the failing stems from an inability to recognize that the power of synergy requires a holistic understanding of where people are positioned." Like Suri, South sees the antidote in a greater respect for spiritual wisdom that stems from a recognition of the spiritual dimension in synergy.

"What we need is a message of hope, a message of reconciliation, the message of forgiveness, love, and unity," he explains. "It requires an acknowledgment of the divine. What is the divine? Another person, Being, community. It's in the room, in the fire, in the bush out there. You have to ask, where are the people standing behind you, beside you, in front of you? Who is directing your steps? Wholeness—our desire to feel whole—comes from a disconnection with the divine. And we need to see ourselves in relationship to others in our community to experience synergy."

Conclusion

Can synergy be created within the Western institutional context, or will any attempt at synergy inevitably be sidelined by the financial and corporate incentives that govern all institutions that operate within a capitalist system born out of colonialism? Will the abiding synergy of the grassroots always be co-opted by the technocratic synergy of the boardroom? In their own ways, Jackson, Suri, and South have all seen how easily attempts to practice true synergy can crack under the pressure to conform to structures that are hostile to any attempt to operate counter-culturally. But each of them also affirms the importance of continuing to push forward in spite of the headwinds.

Stepped Care 2.0 is one attempt to think through how the strands of synergy—truth and Reconciliation, empathy, growth orientation, flexibility, interconnectedness, awareness, contextualization, creativity, diplomatic disruption, and agency—can be operationalized in the context of mental healthcare institutions. Rather than simply being another model for mental healthcare, SC2.0 was designed to create the conditions in which synergy can flow through organizations, creating space in which spirituality and emancipation can be reintegrated into our thinking about what it means to heal. One of the reasons we are looking so closely at the challenges involved in building synergy in an institutional context is due to the fact that SC2.0 has too often been understood simply as an update on business as usual. The embedded logic of institutions exerts a powerful drag on even the most radical attempts to enact change.

It would be naïve to suggest that building a synergistic institution is simply a matter of developing the right policies and practices. But it would also be wrong to argue that institutions can never operate synergistically—as Jackson and South both argued, institutions can foster synergy under the right conditions. More to the point, however, is the fact that, for mental health practitioners operating in Western societies in the twenty-first century, institutions are unavoidable. Perhaps the clearest lesson that should be taken away from these three stories is that operating within an institution will always create a set of risks for the synergy-minded, but with good leadership, good facilitation, and good coaching, these risks are not insurmountable. Ultimately, whether synergy succeeds or fails depends on the willingness of each constitutive part to continue to take the risk of opening themselves up to the radical possibilities that come when the barriers are lowered, trust is established, and the healing energy is allowed to flow.

Reflection

Peter Cornish

Throughout modern history, institutions have not typically welcomed synergy as we describe it. Organizational structures and policies are often aimed at directing control over resources and people. Synergy is a *power-with* concept. Just as synergy existed in the universe long before humans began to exercise dominance over the world around them, synergy will outlast our species. In his book, *The World Without Us*, Alan Weisman imagines the fate of the world should humans cease to exist. Applying principles of physics and ecology, he predicts that while our species' imprint on the world would still be detectable in the distant future to visiting alien archeologists, the planet would soon recover, quickly burying all of our flimsy industrial creations and carnage. Ironically, this macabre realization brings me a feeling of hope. The fact that synergistic ecosystems would prevail, whether civilizations rise and/or fall, whether institutions dominate or crumble, is oddly comforting to me.

I am, perhaps, a tragic optimist. Andrew Boyd, the author of *I Want a Better Catastrophe*, writes about climate doom with gallows humor and a broken heart. The supplementary-narrated flowchart accompanying his book, available for free on his website, is a 52-min, fast-paced, tongue-in-cheek Socratic descent into impossible choices. It is conundrums on steroids with sprinklings of wit, absurdity, humility, grief, despair, and irony.

How can we channel the planet-rattling power of these conundrums? Not without a mix of spirituality, humility, and, yes, some playfulness can we face this threat to our very existence. From a psychological perspective, humor is considered a mature defense mechanism. Humor helps us to cope with stressors in healthy and constructive ways. It allows us to express uncomfortable or distressing feelings in a socially acceptable way, thereby enhancing interpersonal relationships. Humor can help us maintain perspective, ease tension, and approach challenges with a more positive attitude. But as I read Boyd's book and listen to his narrated flowchart, the gallows humor is a meager consolation.

We are facing grief on a scale never before imagined. Grief is the purest form of love. I learned this 25 years ago after my mother died in a single-vehicle accident. My mother was not demonstrably affectionate, so expressions of love were less obvious. It was only in mourning that I recognized the depth of my love for her and hers for me. This was a wonderful discovery, bringing me closer to her in death than I was to her in life. Death can be a beautiful thing. This realization meant that I did not regret what was missing while she was alive. A snippet of dialogue in a novel I am reading now comes to mind, "It's not about getting anything back. It's about having what is right here, right now, and not squandering what remains" (Callahan, 2023). I cannot go back and repair flaws in my relationship with my mother, but I can engage with my memories of her, the land that she loved, and all that is connected with it.

Coming to terms with our mortality, the reality of our flaws, and our limited capacities to relate to one another and the planet we live on can ironically prompt us to take more responsibility in the moment. Existential psychotherapists Viktor Frankl and Irvin Yalom urge us to accept the reality of death, as difficult as this is, so that we may live meaningfully and purposefully with the time that we have. Like Boyd, I feel heartbroken by the prospect of our species' inevitable demise. However, I am buoyed by the possibility of a more graceful exit than predicted by our current socioeconomic trajectory.

The stories presented in this chapter and the chapters that follow draw upon a life force that is synergy. There are many names for this life force in spiritual and scientific practices: Qi (Chi) in Chinese medicine and philosophy, Prana in Hindu and yogic traditions, or biomagnetism in modern science. What is the life force that drove Kareem's allegiance to the team over his own personal ambitions? From where does Genée's passion and commitment to equity and social justice arise in a world stacked against people of color? What propels Raghav to resist institutional brainwashing that disconnects us from all things spiritual?

The answer, I think, is what my mentor, Dick Katz, calls boiling energy (Katz, 1984). It is something much bigger and more powerful than our institutions, and it will surely outlast us. It is in the beauty and wonder that continues to bloom outside our windows despite hundreds of years of industrialization, colonialism, and soulless institution building that fool us with a false promise of agency over the universe. Hope is alive for me when I draw on the life force in the rocks, trees, lakes, rivers, deserts, storms, and clouds. It is alive when I see my fellow humans tapping into, respecting, and celebrating that which is larger than themselves.

Dick Katz's story is fascinating. As a boy, he had a deep sense of spirituality. He shares stories of vivid childhood experiences, like flying through his house as if he were an airplane. He explained to me that he "really was flying," not pretending. Perhaps it was spiritual, which is why it is hard to explain, but it was clear even then that he was tuned into something beyond the ordinary.

When Dick went to university, that sense of "something more" stayed with him. He vigorously pursued yoga, meditation, and other healing practices, eager to explore these yearnings further. At Yale, he soared to the top of his class. Then, at Harvard, he studied under larger-than-life figures. Timothy Leary and Richard

Alpert were exploring psychedelics, while Abraham Maslow was focusing on human potential without drugs. They were pushing the boundaries of elite institutions. But Dick was charting his own course. While Leary and Alpert were captivated by psychedelics and Maslow engrossed by human potential and personal development, Dick was fascinated with spiritual growth aimed at healing and serving others. He saw spirituality as something evolving and communal, not just personal. Leary's mantra was "turn on, tune in, drop out," and Maslow emphasized peak experiences, while Alpert (who later became Baba Ram Dass) explored Eastern mysticism. Dick appreciated their pioneering spirit, but sensed they were missing something crucial: spirituality as a communal and healing force rather than just an individual journey.

Dick taught at Harvard for over 20 years, sharing his insights into spiritual healing from his global experiences. His book, *Boiling Energy*, inspired me early in my career. Dick respected how the Ju/'hoansi's approach to healing in Botswana reflected how humanity lived for most of its history. Their communal healing dance exemplifies synergy in action, highlighting spirituality's role in community and healing.

What struck me most was something he shared with me soon after we first met 30 years ago. The Elders implored him to "Tell our story to your people." Stories have long been my truth. As a transfer student during my undergraduate years, I learned I had used up all my electives. I was told if I wanted to graduate with a psychology degree on time, every course from that point on would have to be a psychology course. At the picturesque campus on a fast-flowing river in rolling green hills, 150 kilometers northeast of Toronto, psychology was considered a hard science. There was not room for broad inquiry. I soon got frustrated and took a year in liberal arts. There, I discovered my love for literature, particularly fiction. Since then, I have sought balance in narrative versus historical truth. Roland Barthes, a French literary theorist and semiotician, examined the nature of storytelling and truth in his essay "The Death of the Author" (1967). While primarily focused on literature, Barthes' ideas challenged the notion of objective, authoritative truth by highlighting the role of interpretation and readerly engagement in constructing meaning from texts. His work contributed to broader discussions about the subjective nature of truth and the multiplicity of interpretations. This multiplicity of interpretations is core to synergy. Science, literature, and art speak truth in human understanding of the universe. Stepped Care 2.0 opens the door to all forms of truth in healing, recovery, and well-being.

Given that positivistic methods have dominated mental health theory and practices for at least 50 years, this volume on synergy seeks balance by reincorporating older traditions that espouse narrative, philosophical, and spiritual truths. It moves beyond the politics, policies, industries, technologies, and institutions that tend to restrict synergy, obscuring windows to the souls of people in community and distracting us from deeper sources of mental and spiritual well-being.

Sometimes, when I share a striking theme or passage in a novel with those who do not share my love for fiction, I am confronted with the question, "But is it real? Is it a true story?" When I reply that it is fiction, they quickly lose interest, perhaps valuing only that which rings true to them. I believe that we are our stories. Patti Callahan, the author of the novel that I quoted above, reminds us that "we are a myth-making people. It is how we make meaning of the meaningless and sense of the senseless" (Callahan, 2023). Unfortunately, large established institutions representing organized religion, profit-driven industries, or elite universities funded mainly through the STEM disciplines maintain their status by dismissing ancient popular wisdom, folklore, magic, and myth as cultural accouterments. In this volume, we contend that psychological, physical, and spiritual healing is only possible through the inclusion of myths, stories, songs, dances, or rituals that allow people to connect with one another as well as something much larger than us. Synergy is the sustainable energy that guides healing and recovery. The SC2.0 framework is nothing special without it.

References

Barthes, R. (1967). The death of the author. In *Image, music, text* (pp. 142–148). Fontana.

Callahan, P. (2023). *The secret book of flora lea: A novel.* Simon & Schuster.

Katz, R. (1984). *Boiling energy: Community healing among the Kalahari Kung*. Harvard University Press.

Chapter 4
The Gaslighting of a Profession: DEI Training in the Mental Health Profession

Catie Greene

> *To educate as the practice of freedom is a way of teaching that anyone can learn. That learning process comes easiest to those of us who teach who also believe that there is an aspect of our vocation that is sacred; who believe that our work is not merely to share information but to share in the intellectual and spiritual growth of our students. To teach in a manner that respects and cares for the souls of our students is essential if we are to provide the necessary conditions where learning can most deeply and intimately begin.*
>
> —bell hooks, 1994, p. 13, *Teaching to Transgress*

Who Are the Diviners?

As a counselor educator and supervisor, every semester that I teach a field experience course, I end our first class with the following lesson. I begin by asking my students, "What are the steps that one takes to become a mental health professional healer in our society?"

Students:	*"Pass a licensure exam?"*
Me:	*How?*
Students:	*By studying?*
Me:	*Studying what?*
Students:	*[List off studying materials for their licensure exams]*
Me:	*Okay, so you study written materials.*
Students:	*Yes.*
Me:	*Okay, what does the exam look like?*
Students:	*Multiple choice and some written sections.*

C. Greene (✉)
Stepped Care Solutions, Mount Pearl, NL, Canada
e-mail: catie.greene@steppedcaresolutions.com

P. Cornish, G. Berry (eds.), *Synergy*,
https://doi.org/10.1007/978-3-031-88493-1_4

Me: *What else does it look like? Where are you? Who's with you?*
Students: *A few hours. In an exam room. Maybe a proctor. On a computer.*
Me: *Okay, what do you have to do to get to that point to take the exam?*
Students: *First you need to graduate from an undergraduate program. You also need to graduate from a master's or doctoral program in counseling, social work, or psychology.*
Me: *Okay, how do you graduate?*
Students: *You need to earn X number of credits to graduate from an undergraduate program. And X number of credits to graduate from a graduate program. And do a field experience.*
Me: *How do you pass your field experience?*
Students: *You have to get X number of supervision and clinical hours. You have to pass your supervisor evaluations.*
Me: *How?*
Students: *You earn X score on your final evaluation completed by your supervisors.*
Me: *So, in our society, one earns a certain number of credits by passing a certain number of courses, over a certain number of years. One then completes a certain number of clinical hours under supervision from another professional, and if their skills are evaluated to be satisfactory, they earn a certain score on an evaluation and are able to pass their field experience. Then they complete a written exam, usually alone in, like, a Pearson exam room?*
Students: *Yes.*
Me: *Okay. How are you chosen to do so?*
Students: *Chosen?*
Me: *Like, do you get selected to train to become a healer in some way?*
Students: *Yes, during the graduate interview process [now they're looking at me as if to ask: "Weren't you there?"]*
Me: *But before that. Before you submitted your applications for selection. How are you chosen to do so?*
Students: *...we chose to, ourselves?*
Me: *Okay. Got it. So, as I go through the next few slides, I want you to compare what you've just shared with me to the requirements of a South African Diviner, or traditional healer.*

I then share the following segments from Mokgobi et al.'s (2014) paper:

> For certain categories of traditional African healers such as diviners, training is a formal and meticulous process that *can take between months and years depending on how fast the trainee learns the trade* (Peek, 1991).

I emphasize: *Between months and years, depending on the learner. Not X number of credits, X number of courses, X number of years, standardized across thousands of learners.*

> To become a traditional healer, a special calling from the ancestors is required. The authenticity of such callings is verified by a diviner who advises on who should undergo training at an appropriate trainer.

Chosen by the ancestors, verified by an advising diviner. Not self-selected.

> During training, the trainee is required to live with their trainer, the trainer's family, and other trainees, and is therefore constantly observed by the trainer (Rudnick, 2002). During the training process, trainees receive instructions on a variety of aspects such as different medicinal plants and animal extracts to use, interpreting bones, dream analysis, communicating with the ancestors and different illnesses and how to treat them.

Living with their trainer. Evaluated holistically. Observed both while practicing and not practicing.

> Once the training is completed, the trainee is taken to a river where final rituals are performed at a ceremony in the presence of community members, called "go ja ntwase."

At the river. In front of their entire community. Not alone, at a computer, in a Pearson exam room.

> This ceremony is a form of an assessment to test if the trainee has mastered the trade and can be allowed to practice as a traditional healer (Mutwa, 2003). If it happens that the trainee fails this assessment, the training may be extended by some more months.

The community that you will be healing helps decide if you are ready to be their healer. Not the number of items you answer correctly or your numerical exam score. And if you're not ready, you return to training for an undetermined amount of time until you are ready. You don't return to studying for the exam, but to training, before being invited back to the exam.

> Not every qualified traditional healer is qualified to train prospective traditional healers. Training of traditional healers is a specialty and yet another calling, in addition to simply being a healer. A traditional healer has to be called to become a trainer of other future healers.

To be a supervisor/mentor to future diviners requires another special calling from the ancestors, not an employer requirement.

I then summarize: *Becoming a professional counselor/social worker/psychologist means self-selecting as a healer in our Western society, trained by other* self-selected *healers.*

At this point, I display a selection of photos of current civil rights issues in our society, and pose the questions: *Are you personally, spiritually, and professionally prepared to be a* self-selected *healer in this society? Are you doing the work? Are you prepared to do the work?*

I know that most counselors in training are already undergoing near-daily internal battles: *Am I good enough? Did I make the right career choice? Do my peers think I'm good? Does my professor like me? Will my clients like me?* I do not introduce this comparison to induce more fear, self-doubt, or anxiety. I introduce it because I think it is important that we acknowledge the reality of our choice to become a healer. I want counseling trainees to feel this choice deeply: This sacred choice to be a healer, right now, in this particular society with so much cultural pain and need.

We then usually discuss the "gatekeeping" measures that are in place to ensure a trainee's successful and ethical application of basic counseling skills and techniques. However, I tell my students truthfully that these measures are insufficient. Oftentimes, our approaches to such measurement are flawed in favor of good students versus good healers (Brear & Dorrian, 2010; Guenette & Marshall, 2008; Jacobs et al., 2011; Sowbel, 2012). Realistically, they will likely graduate and tout a professional license with no co-signature from any member of the community they will serve in the future, let alone their ancestors. As there is no spiritual component of such evaluations (such as those that have been inherent in the training of Indigenous healers for centuries), they can do so with no real intrinsic desire or ability to care for other humans, community, justice, and human diversity. Islamic psychologist Bendeck Sotillos (2022) critiques mainstream therapy training for its absence of attention to the counselor's spiritual growth; "we can only provide treatment to others to the degree that we ourselves have faced our own trials, both as they apply to everyday life and to the spiritual path. Again, the role of the therapist in this context radically differs from what is expected in mainstream psychology" (p. 414).

South African diviner training is not unique in its emphasis on the spirit or the soul of the healer-in-training. In the majority of communities across the globe, particularly in communities Indigenous to their land, training to become a healer involves attention to the spiritual realm (Ermine, 1995; Guenette & Marshall, 2008; Moodley et al., 2008). However, in mainstream training programs, individuals can become healers with straight A's and a glowing licensure exam score, but without having gained much awareness of who they are spiritually as human beings, their interconnectedness to the world, how their socialization has shaped them, and how, exactly, they are impacting others in their attempts at healing.

I do not share this with my students to scare them, but to communicate the importance of the individual responsibility to understand oneself and one's own cultural programming, to love humans enough to show up with healing energy in times of sociocultural conflict, and to set the tone in a classroom (a space which could not be more devoid of spirituality) that this work is sacred work. It is then my job, as bell hooks puts it, to care for their souls and help them elicit their own unique strengths as we share a semester together. The hope is that, if they choose to do this work, they will translate that depth of care to the people who entrust them with their souls. These are messages that my own mentors, such as Dr. Gillian Berry, who embodies synergy in everyday life, have impressed upon me. They have pushed me to see beyond the ways we currently approach well-being practices and integrate diversity, equity, and inclusion (DEI) in the mental health professions.

Who Am I and What Is This Chapter About?

My name is Catie Greene. I am a licensed professional counselor and counselor educator, and I have spent my career straddling academia and clinical practice—at times, as a full-time clinician and adjunct instructor, and, at others, as a full-time lecturer and part-time clinician. I first learned about the Stepped Care 2.0 (SC2.0) model when I worked for a college counseling center that implemented it just a year before I joined. It was there that I met my mentor, Gillian, and was able to work with a team of empowered, diverse, and creative clinicians under her unique guidance. I would later learn that this counseling center had become an early adopter of the SC2.0 model after Gillian and other members of leadership attended a less-than-one-hour-long presentation by Dr. Peter Cornish. I was told the model resonated with Gillian's own background, training, and experiences of being a diplomatic disruptor in the current mental health care system, and that Peter and Gillian immediately developed a strong cooperative relationship after that one interaction. After several years at the counseling center, Gillian transitioned to work for Stepped Care Solutions (SCS), a company with the lofty mission of "wellbeing everywhere," and it was not long before I followed her on her next adventure and joined the synergy team at SCS. Originally intended to be the SCS DEI team, the synergy team, from its inception, has focused on aligning the SC2.0 model with an expansive philosophy of healing we believe is necessary for true well-being everywhere.

Working on the synergy team has only introduced me to more mentors. For example, I had the privilege to "sit at the feet" (as Gillian would say) of Dr. Richard "Dick" Katz at the retreat described in the next chapter by André Forget. It is amazing how the power of mentorship was so palpable in just a few days together, and how seamlessly all of us fell into multicultural familial generations of grandparents, parents/aunties/uncles, and children/siblings/cousins in our interactions. My place in the generations at this retreat was securely in the "child" generation, as my sibling/cousin colleagues and I all but ate up the words and mannerisms of our elders. I have so much more to learn. The theme of mentorship was so powerful that my SCS colleagues, Kareem South and Sarah Harte, and I, remarked how uncannily Dick's warm face resembled those of the mentors he honored at the beginning of his talk, though they were from different racial backgrounds, cultures, and continents.

I have been invited to write this chapter as a student of the concept of synergy, and to provide some historical context as to why I believe the success of SCS in mental health system transformation is reliant on its willingness to go beyond the ways our field has traditionally attempted to incorporate DEI. Just as Dick's mentors' faces could be physically seen through his own as he spoke of them, my words may be my own, but I believe they reflect my mentors speaking through me. Or, at the very least, they reflect my current understanding of my mentors' messages synthesized with my own experiences, and what has resonated authentically with me. In addition to academic references, I share my own experiences as a counselor educator and supervisor, and as a human whose healing has been transformed by the difference of synergy in the classroom, in the workplace, and in my broader life. At times, full quotes of another's words will be woven in, such as the late bell hooks' or Robin Wall Kimmerer's. Though there are differences in their messaging, cultures, publishing pathways, and professions, the beauty of beginning to see the world synergistically, for me, has been to feel the interconnections between seemingly different human beings, all contributing to our collective healing.

The Gaslighting of a Profession

> *I didn't think about it at the time, but it was happening all over again, an echo of my grandfather's first day at school, when he was ordered to leave everything—language, culture, family—behind. The professor made me doubt where I came from, what I knew, and claimed that his was the right way to think. Only he didn't cut my hair off.*
>
> —Wall Kimmerer, 2015, p. 41 *Braiding Sweetgrass*

The synergy team members and co-authors of this book and I decided to title this chapter "The Gaslighting of a Profession" following a conversation we had about a recent facilitated workshop at SCS. As a synergy team, we facilitate sessions for staff to engage with concepts related to diversity, equity, and inclusion (DEI) in well-being practices. In this particular session, we were discussing the concept of cultural appropriation in the context of SCS's mission of global well-being. We engaged questions through the lens of synergy such as: *When is it that we are truly sharing ways of knowing and healing resources to create "a sum greater than its separate parts"? When is it inauthentic, stealing, obfuscating, furthering cultural domination, and reducing healing potential? When is avoidance of being perceived as perpetuating cultural appropriation actually performative? When might the impact of that performance be more dehumanizing than culturally humble or respectful?*

We used a recent example in which some staff expressed fears that others were perpetuating cultural appropriation by participating in a drumming circle performed by White French Canadians who were trained by African drummers. Staff vulnerably shared their fears of speaking about other cultures' healing practices when their own culture is more privileged, wearing jewelry made by individuals from groups that were less privileged relative to them, and their confusion around how they should feel about the drumming circle the staff experienced. On the one hand, some staff did not want to support a White group of drummers who were privileged in comparison to the African tribal drummers who taught them their craft, and on the other, they were not confident that they understood the context well enough to know what was "okay" or respectful. The dialogue was tentative, marked by uncertainty about how to connect and participate with cultures outside one's own while respecting the realities of power, privilege, and cultural domination. It was also quite intellectual, focusing mainly on how to navigate the many competing ethical principles at play—until, that is, one staff member in the margins of the Zoom chat typed:

> You know, honestly, I was really feeling the drumming and had a great time! I didn't even think of all of these issues. I guess it's because I don't really know these theories and terms, like cultural appropriation, and things like that.

This staff member was the only person in the discussion who was actually born and raised in an African country.

It was the last sentence that really hit me: "I guess it's because I don't really know these theories or terms." Despite having a different feeling than others expressed about the drumming circle, this staff member second-guessed if that might be because of her lack of familiarity with this particular DEI concept. At this moment, I was reminded of a host of similar examples I have witnessed in my career when well-intended discussions around DEI and well-being create a sense of doubt and disconnection in the very people whose feelings and perspectives it aims to be inclusive of.

I remembered the time one of my Nigerian-American graduate students spoke to me after class and told me her goal was to "bring mental health to Nigeria" following her graduate degree program. As if the country of Nigeria had no mental health to speak of. Or, when discussing my work, I have used the acronym "DEI" in favor of more everyday words and someone in my community—whose actions exemplify all three—says, "What's that?" and I realize I have put myself in a position of teacher in a conversation when I am really the one with so much to learn.

It also reminded me of when I recently asked the healer who helped facilitate our talking circle at our synergy retreat (described in the next chapter) if she would be interested in participating in a podcast on one-at-a-time thinking in therapy. Despite the essence of one-at-a-time thinking being central to Indigenous healing practices, she said, "I'm not sure I have enough to offer" because she was not familiar with the very new, specialized phrase "one-at-a-time therapy." Talk about cultural appropriation.

When concepts of Western healing and DEI are so inaccessible that people start to question their body's response to the beat of a drum, or the healing lessons taught to them by their Elders as having much to offer, or the very existence of any mental health healing resources in their home country, we must admit that there is something wrong with our delivery. As Wall Kimmerer (2015) said as she compared her experience in college with her grandfather's experience of being forced to attend the well-documented traumatic, abusive residential schools (Bombay et al., 2014) designed to strip him of his Native culture and practices; "The professor made me doubt where I came from, what I knew, and claimed that his was the right way to think. Only he didn't cut my hair off."

In fact, many of the Western healing professions have done just that. This was most recently apparent during the global civil unrest and racial awakening of 2020–2021, when we saw a trend of mental health professionals admitting to wrongdoing in regard to multiculturalism and DEI. For example, the American Psychological Association (2015) stated that it had

> failed in its role leading the discipline of psychology, was complicit in contributing to systemic inequities, and hurt many through racism, racial discrimination, and denigration of people of color, thereby falling short on its mission to benefit society and improve lives. APA is profoundly sorry, accepts responsibility for, and owns the actions and inactions of APA itself, the discipline of psychology, and individual psychologists who stood as leaders for the organization and field.

In another statement, the American Psychiatric Association (2021) promised to address the racism that exists in the psychiatry industry, saying that

> The APA is beginning the process of making amends for both the direct and indirect acts of racism in psychiatry. The APA Board of Trustees (BOT) apologizes to its members, patients, their families, and the public for enabling discriminatory and prejudicial actions within the APA and racist practices in psychiatric treatment for Black, Indigenous and People of Color

> (BIPOC). The APA is committed to identifying, understanding, and rectifying our past injustices, as well as developing anti-racist policies that promote equity in mental health for all.

The National Association of Social Workers (2021) also stated:

> As the nation looks at its long, cruel history of systemic racism, the National Association of Social Workers (NASW) acknowledges that our profession and this association have not always lived up to our mission of pursuing social justice for all. NASW apologizes for supporting policies and activities that have harmed people of color.

The American Counseling Association did not issue such an apology nor acknowledge its participation in perpetuating racism in mental health, but it did make the following "anti-racism statement" in 2020:

> We stand in solidarity with our Black siblings in denouncing the historical legacy and destruction caused by institutionalized racism and violence against Black people, perpetuated at the hands of law enforcement, the hatred bred of White supremacy, the deafening silence of dehumanizing and complicit inaction to address these systemic ills within our society. As counselors, we listen, we empathize, and agree with protestors that when absolute justice is established, peace will follow. Enough is enough, we cannot continue to watch fellow Black Americans being murdered, as the very life force is suffocated out of them. [...] Our stance is: Black Lives Matter. We have a moral and professional obligation to deconstruct institutions which have historically been designed to benefit White America. These systems must be dismantled in order to level the playing field for Black communities. Allyship is not enough. We strive to create liberated spaces in the fight against White supremacy and the dehumanization of Black people. The burden of transgenerational trauma should not be shouldered by Black Americans even though they have remained resilient.

Despite these public apologies and pledges of anti-racism, the proposed steps forward generally speak to changes that do not threaten the foundation of the mental health care system: its commitment to privileging a Western philosophy of healing in favor of a philosophy that respects and believes in the importance of respecting a diversity of ways of knowing for collective healing (Bendeck Sotillos, 2022).

Oftentimes, DEI training initiatives emerge from the mess of this flawed ideological foundation when long-standing calls for reform are briefly heard during times of civil unrest (Anand & Winters, 2008; Celinska & Swazo, 2016; Engram & Mayer, 2023; Leahy et al., 2009; Paluck & Green, 2009; Singh et al., 2020; Smith & Trimble, 2016). However, these initiatives sit atop unquestioned assumptions about the "right" way to heal and practice healing (Bendeck Sotillos, 2022). The American Psychological Association defines *gaslighting* as "to manipulate another person into doubting their perceptions, experiences, or understanding of events" (APA, 2023). Ironically, in a gaslighting effect, the very institutions that have been socialized and socialize others from a perspective of domination, then turn around to teach DEI principles through their philosophical lens. Imagine the perpetrator of your trauma undergoing no meaningful healing themselves and trying to teach you how to heal from the wounds they inflicted on you?

In 2021, in the midst of these apologies and reckonings, a time when the field was arguably ripe for deep, wide-reaching changes, the *Journal of Mental Health Counseling* published Branco and Jones' article "Supporting Black, Indigenous, and People of Color counselors: Considerations for counselor skills training and practice." Despite the firm acknowledgment that; "Client racism toward BIPOC counselors is a systemic problem and therefore requires a systemic response, to include counselor educator and clinical supervisor efforts, to avoid BIPOC counselors shouldering the weight of racism and microaggressions in isolation" (p. 282), I think the recommended skills and strategies to "support Black and Indigenous and People of Color Counselors" fall short of supporting and training such counselors as human beings with souls and spirits.

Before sharing my critique of this article through the lens of synergy, I want to acknowledge that I do not know what possible humanity may have been stripped from this article during its publication process. I do not know how radically empowering of BIPOC students, as dynamic people with souls and spirits, the authors' ideas would be if given the freedom of expression rarely found in academic journals. All that I can speak to is the ways in which what was published appears to me to further the status quo.

For example, though reportedly consistent with Critical Race Theory (CRT) and *microinterventions* "response strategies" created "in an effort to dismantle the potency of a microaggressive encounter" (p. 284), the authors offer a clinical vignette of a microaggressive client and suggested responses from the counselor in training to be aligned with counseling skills and ethical standards. The following clinical vignette is what is offered as an example that is supposed to be consistent with addressing "the clinical training needs of minoritized and marginalized counselors":

Client:	I am wondering something. Are you legal to work in the United States? I ask because I do not want our work with you to interfere with my security clearance.
Counselor:	You are interested to know if I have documentation to legally work in the United States, as you want to preserve your security clearance. This sounds quite important to you.
Counselor:	What concerns you about my legal status?
Client:	Yes, that is accurate. Well, you look Hispanic, your last name sounds that way, and I notice the art in your office looks that way too, so I am just checking that before we get started.
Counselor:	I appreciate your honesty and observation of the artwork. I welcome questions about these issues as we build our counseling relationship.
Counselor:	And to go back to your question, I do identify as a member of the Latinx community. And like many members of the community, I have legal status to work in the United States, as I am a U.S. citizen (Branco & Jones, 2021, p. 287).

The counselor's responses in the above vignette are the authors' recommendations for how a BIPOC counselor should respond to the client in order to uphold the counseling profession's ethical and clinical standards. In order to support the BIPOC counselor, the article recommends that supervisors should help the counselor in such scenarios to reflect on questions surrounding how the microaggression made them feel, what self-care and wellness strategies they have for support, and when they have experienced similar microaggressions before. Purportedly, these counseling skills and intervention recommendations were developed to combat the cultural supremacy inherent in traditional helping skills training. The authors note that "BIPOC counselors, who enter the training experience and profession with lived racialized experiences, may benefit from clinical skills instruction, rather than replicating the current White hegemonic counselor education framework (Branco & Jones, 2021, p. 282; Fuentes et al., 2020; Spanierman & Smith, 2017).

Yet, it is unclear to me how the "White hegemonic" counseling training framework is being disrupted here. Instead, it appears that it is being strengthened by the claim that it can be "adapted," yet adhere to the same basic counseling skills and counseling ethical standards of the profession. In my perspective, it disregards the counselor's humanity and asks them to do the same in favor of upholding a false sense of a therapeutic relationship for the client. Given the helping professions' public apologies and pledges to anti-racism issued in the same year as the publication of this manuscript (American Counseling Association, 2020; American Psychiatric Association, 2021; American Psychological Association, 2015; National Association of Social Workers, 2021), this article is disappointing to me, but not surprising. The profession's commitment to finding increasingly complex and inaccessible ways of communicating its intentions to be more diverse, equitable, and inclusive is only matched by its commitment to the status quo. As the Association of Black Psychologists (2021) articulated in its response to the American Psychological Association's (2015) apology:

> The American Psychological Association (APA) cannot simply dismiss the full history of Euro-American psychology that is rooted in a legacy of the pathology of Whiteness with a simple apology and questionable claims to now combat racism, oppression, and white hegemony. The APA, in fact, has played a large hand in the oppression of the Black community in education, health, housing, the media, the work sector, criminal justice, and practically all domains of life necessary to thriving and optimal well-being. Borrowing again from this transformative concept of Sankofa, we also know that "the white man will try to satisfy us with symbolic victories rather than economic equity and real justice," underscoring that while the apology and resolutions may appear genuine, it is far too late for it to mean anything to Black America and sorely lacks several opportunities for real and meaningful apology and atonement, which are inextricably linked to restitution and restoration.

If Branco and Jones' article is meant to center the needs of BIPOC counselors in training, why does not the reader get to know anything about the human on the receiving end of the microaggression? We have no idea how the microaggression affected them, what thoughts ran through their mind, and what went into their clinical decision-making with each response they provided. Instead, they are rendered as separate, to be explored in supervision or "bracketed" to reduce clinical bias, all while the reader is presented with prescribed responses. We are left with no idea whether the responses are, in fact, ones that maintain the therapeutic relationship if we do not know how the client's words or assumptions uniquely affect the counselor in training. Microaggressions often stun individuals on the receiving end, resulting in fawning or people-pleasing to maintain the semblance of connection (Walker, 2008). What strategies of disconnection might this counselor in training be exhibiting to protect themselves in this clinical vignette in order to provide those prescribed responses? How could they be supported to make use of the feelings that do come up for them to then inform a clinical intervention that honors the humanity of all members of the therapeutic relationship? How might that be able to be reflected in a vignette in an article attempting to help counselor educators and supervisors support a student?

In *How Therapy Helps When the Culture Hurts*, Relational Cultural Theory scholar Maureen Walker (2008) offers the following to the reductive question of "What is a therapist to do?" in the context of cultural disconnection:

> Again, the apparent question is not always the compelling question. The compelling question before us was not about disclosure or nondisclosure, but rather how to gain the clarity and courage to embrace a potentially healing and empowering conflict. It was helpful for me to pause long enough to connect with the feelings-thoughts that emerged and morphed as we continued the conversation. It is usually during these moments of noticing that I can connect with the support and holding I have received from supervisors and mentors over the years: people like Irene Stiver who always reminded us that in moments of confusion or complexity, we can usually find "one true thing" to say that will (1) keep us connected to the relationship and (2) allow other truths to emerge (p. 7).

Now, returning to the Branco and Jones vignette with this in mind, imagine instead that we learn that Gabriela is a second-year multilingual counselor in training. She was born in the United States and raised between Miami and Venezuela. She is one of a few Latina students in her clinical training program and the only one whose native language is Spanish. Though Gabriela always regarded her multilingualism as a strength, she has not felt this way throughout her time in her program. Instead, Gabriela notices she becomes more cognizant of her accent in class and in sessions, often resulting in her expressing much less than she wants to share. That is, until her university supervisor spends time in their first group supervision session in class encouraging classmates to share about their cultures and identities. When Gabriela apologized that English was not her native language, her university supervisor encouraged her to consider saying instead that she was "multilingual" and to share which languages she did speak. This was the first of many reframes that encouraged Gabriela to take up more and more space in class. Even though she and her university supervisor have major differences in identities and cultures, she found that her supervisor's authenticity and creativity made her curious about how she can do the same as she progresses in her own counselor identity. In their individual supervision sessions, she always feels challenged, encouraged, and that her experiences and identities were a strength to the program. She still feels very nervous about being a new counselor but trusts she has the support of her supervisor and classroom environment to navigate these challenges.

She has only conducted a few sessions on her own when she meets Martin, a third-generation, White Italian-American man, who comes to therapy for the very first time. As Gabriela welcomes Martin to his first session and begins to ask him about his goals, he quickly cuts her off to say:

> **Martin:** I am wondering something. Are you legal to work in the United States? I ask because I do not want our work with you to interfere with my security clearance.

Gabriela feels her chest become warm and her throat tighten. She quickly says "Yes, of course, I am" more quickly than she would have liked to. She briefly loses track of her previous question as the room falls into an awkward silence. She does not like how immediately she felt flustered by this, but that feeling was quickly replaced by the warmth and confidence she knows she will feel once she shares this interaction with her university supervisor, who she knows she will not even need to explain how this was a microaggressive interaction. She feels safe knowing she can simply do her best right now, that her best will be enough, and that she can process the interaction after the session. She thinks about what her university supervisor would encourage her to do right now, and can hear them saying; "You belong here, Gabriela. Your experiences are valuable in this field. When you don't know what to do or say in the face of a difficult clinical connection, know that you can always say 'one true thing' (Walker, 2008) *that could keep you in therapeutic connection without sacrificing your authenticity." She knows she "should" ask him why he asked her that in the form of a question or reflection, but realizes she is really not feeling very curious right now to hear more from Martin about why he asked that question. She knows she will be in time, but not right now. And feigning curiosity right now would move her away from connection. She is able, instead, to feel his nervous energy. And the more she notices it, the more she finds herself wanting to move toward him. To understand what actually brought him here.*

Gabriela: I can imagine it might be really nerve-wracking to be in this space for the very first time.

Martin: (tersely) Yeah.

Gabriela: I understand that. And it may take some time to trust if I can be helpful to you. Sometimes, though, the only way to find out is to try. And I have to trust you that you will know if it is working for you.

Martin: (nods)

Gabriela: I was beginning to ask you before what you came here to talk about. But now I am thinking I will rephrase that instead to, what would you most like to happen in our time together today?

I am not suggesting that my vignette above is more "correct" than Branco and Jones'. I believe that if the responses provided in their vignette were authentically the "one true thing" the trainee in question might have said, and the trainee felt empowered and confident in their role as a healer in the process, then those are the responses that make sense. In honesty, when I read their vignette, I cannot help but feel like the counselor is straining themselves to be someone that they are not. To sound like "Susie the social worker." And I feel a sense of dread thinking about one of my own students reading that vignette and thinking that that is how they should sound, instead of reaching within themselves and integrating their training skills to expand that. As Dick Katz said during his lecture at the synergy retreat (2023), "I was trained as a clinical psychologist […] We're trained to interpret, to tell people what's wrong with them, we're trained to be helpful. Which is the worst thing to go for. You don't want to be trained to be helpful. We want to be trained to listen. Behind your credentials as professional helpers is a story. It's not a lesser story. It's *your* story." I am so grateful that I have learned from my mentors that to be a healer is to learn how to be truly congruent with who we are, and expand that healing energy outward. But this is unfortunately not the norm in the way we train therapists today.

As I offer my critique, I also want to note that I hear Dick Katz's words in the back of my mind saying that to dismiss another's approach to this work as less than in the name of synergy is just another form of cultural domination, as true synergy exists when a variety of seemingly disparate healing methods can enhance one another (Personal Communication, October 2023). I know I have a ways to go on my own spiritual journey as a healer, and I am mindful that if my words suggest that *I* have a better way of approaching things than another, this is not synergy. However, this is where the context of cultural domination creates a level of pressure to be perfect when one offers solutions to problems of inequity (see Poser, 2024 for an account of the complicated position Ibram X. Kendi faced following the release of his bestselling book *How to Be an Anti-Racist*). If academic journals published a wide variety of articles on the topics of supporting BIPOC counselors in training and navigating microaggressions in therapy, and Branco and Jones' approach was just one of many, this would create a very different context. Instead, an article like Branco and Jones' exists as one of very few articles ever published on the topic, and therefore I want it to do more than one article could ever do. My disappointment is not with Branco and Jones but the context that situates their approach to supporting BIPOC counselors in training as *the* approach that aligns with ethical and practical standards of counseling, rather than one of many approaches to navigating the challenges inherent in training racially marginalized counselors to offer healing in a racially unwell society. My concern stems from how the approach Branco and Jones speak of is already so privileged in our mainstream healer training, and how long I imagine it will take the same academic journal to publish something on the topic outside of its comfort zone. And the amount of racial trauma it will take to reach another breaking point in which such articles are briefly welcomed.

The Comfort Zone: The Banking Approach to Education

Paolo Freire is famous for his "pedagogy of the oppressed" concept of teaching, which encourages critical thinking, empowerment, and engaged dialogue (Freire, 1972). Freire saw his approach as being in contrast to the traditional Western "banking approach" to education (Freire, 1972). In the banking approach, educators deposit information, and students are simply considered receptacles for the knowledge a teacher holds. In critiquing DEI/multicultural training in the Western helping professional education, Halman et al. (2017) argue that

> To effectively educate psychologists and counselors to become culturally competent, social justice advocacy requires a paradigm shift from traditional teaching methodologies to creative teaching techniques (Chung & Bemak, 2012). Paradoxically, many traditional teaching strategies are contradictory to the aims and goals of multicultural social justice training. The banking concept of education is in direct contrast with the goals of multiculturalism and social justice in that it promotes disempowerment rather than empowerment, and reinforces the idea that educators/teachers have all the knowledge and students are simply empty vessels and subsequently passive recipients of information (Freire, 1972). In the banking approach to education, students are not expected, nor are they given permission, to question, critique, analyze, discuss, assess, or evaluate information. Rather, students are expected to submissively take in and then regurgitate the information that has been given to

> them by their teachers. The banking approach to education promotes conformity, resulting in an unhealthy dependent relationship in which students are reliant on their teachers. This colonial mentality of education creates an unequal balance of power and crushes any form of independent thinking and inquiry, aiming to produce conventional and conforming individuals who simply accept rather than question information deposited into them (Chung & Bemak, 2012).

Most of us involved in the training of future healers do not want to believe that this is our dominant reality (Smith & Ng, 2009). If we search the literature on counseling, psychology, and social work pedagogy, we will find numerous articles, chapters, and entire books describing constructivist, relational, multicultural, and epistemological approaches to counselor training and supervision (Brubaker et al., 2010; Dollarhide et al., 2007; McAuliffe & Erikson, 2010; Stadler et al., 2006). These approaches discourage the banking approach in favor of those that invite and involve the adult learners' perspectives. There are incredible educators doing wonderful work to decolonize counseling, psychology, and social work education. However, the systems in which this training occurs inherently privilege the banking approach to education, such as through large class sizes, curricular requirements, grading policies, and an unquestioned underlying philosophy of healing (Deepak et al., 2015; Goodman et al., 2015). It is no wonder our faculty teaching diversity and social justice courses consistently report lack of effective gatekeeping, institutional support, burnout, and disappointment with the textbooks and curricula they are charged to make do with (Deepak et al., 2015). In my experience as an educator, it often takes a considerable amount of work to justify teaching methods that fall even minorly outside of the prescribed curriculum.

In their book *Synergy, Healing, and Empowerment,* Dr. Richard Katz and Dr. Stephen Murphy-Shigematsu (2017) describe an example involving Paolo Freire that illustrates the type of structural disruption to the status quo that is needed to distinguish between a synergistic learning environment and a non-synergistic learning environment that purports to teach concepts of DEI. In the example, Katz describes the behind-the-scenes decision-making that went into having Paolo Freire as a speaker in his psychology training program. At first, the plan was to have Freire speak to an audience of students and faculty in a lecture hall. With so many students and faculty excited to engage with Freire, some felt the lecture hall approach, despite its banking methodology, was the most "fair." However, this did not feel quite consistent with his engaged approach to pedagogy. They considered limiting attendance to allow for deeper engagement, which presented a host of other issues of equity. Through further discussion, the committee decided to disrupt this traditional approach to "fairness" (Katz refers to it as the "scarcity" approach to the resource of Freire's teaching) and instead have Freire participate in one of the several small working groups made up of students and faculty. While this meant that most attendees would not get to directly engage with Freire, it allowed all members to actively engage with the purpose of his visit and to actively bring their perspectives into dialogue. As a result of the structural disruption to the status quo, attendees ended up helping to create more renewable energy as a group than they would have created as passive recipients in the lecture hall. Instead of considering the attendees as privileged and lucky to have participated, they can be seen as entrusted to sow the seeds of synergy they tapped into by practicing what they learned in their community.

Without the institutional support, willingness, or ability to engage in DEI training creatively and disrupt the status quo, we miss out on such opportunities to expand diverse healing resources. While student resistance to DEI training due to racism, bias, and internalized oppression are well-documented and play a significant role in the challenges of DEI training in the mental health field (Constantine et al., 2008; Goodman, 2011; Watts, 2007), I think that unquestioned Western assumptions about learning and healing also underlie these issues. In fact, they may be even more dangerous, as merely questioning them can result in consequences—despite Freire's insistence that true learning requires such questioning.

The Impact of "Progress" that Stays Within the Comfort Zone

As my students' vocabularies have grown over the years, with terms like decolonization and White fragility becoming more commonplace in the classroom, I think two essential constructs remain significantly underdeveloped. These are the constructs of interconnectedness and humanity. When the soul is not engaged in the learning process of becoming a healer, it follows that such concepts might lag behind vocabulary (Ekpe & Toutant, 2022). I believe this is a direct result of the limitations of the Western approach to DEI training when compared to a more holistic approach. Bettina Love explains this as the difference between an "ally" and a "co-conspirator." "Allies know all the language, [but] to be a co-conspirator… is to take risks for somebody" (Love, 2019).

In the same talk, Dr. Love explains that allies tend to be the folks who have read all of the books and know all of the terms when discussing issues of race and racism, but in a moment of truth, resort to inaction out of fear and avoidance. We also see this play out in the classroom when well-meaning White students are quiet when the class is discussing issues of race. Having been introduced to the concept of privilege and "taking up too much space," they tell me that they are trying to "give space" to others (i.e., their peer Students of Color) during the dialogue. Unsurprisingly, this leaves Students of Color feeling isolated from and observed by their peers rather than engaged with. When they articulate this to their White counterparts, their White peers reveal that they are "afraid to say the wrong thing." In fairness, we often do respond defensively when areas of our privileged identities are exposed (Watt, 2007). Watt and colleagues studied counseling trainee responses to multicultural dialogues and found a range of responses along a continuum that serves to assuage one's own denial, guilt, and anxiety when presented with facts about areas of privilege (Watt, 2007). These defenses do, in fact, take space away from connection and healing—they also become a large part of the instructor's classroom management (Williams et al., 1999; Young & Davis-Russell, 2003). But when counseling trainees feel the only solution to this problem lies in extracting themselves from human connection with their peers, we as teachers and trainers have to reflect on whether or not our approaches to DEI training have offered students a viable alternative. How might we be perpetuating this inaction? The public might be troubled to know just how many trainees have moved through their entire training experience successfully avoiding deep, diverse, and risky human contact in this way.

We also see this fear and avoidance when it comes to treating and teaching counselors who are People of Color as if they are human beings. In the midst of the 2020 racial reckoning in the United States, I attended a faculty meeting that had "discuss current events of racism" listed as an agenda item. One of the department heads asked how we felt we could improve on anti-racism. With the little time we had, the dialogue that emerged focused entirely on strategies for reducing bias in grading and evaluation, and examining whether we were reproducing any unfair grading practices that make it harder for Students of Color to be successful in the program, or more likely to fail. One of the outcomes of the dialogue was that we would more frequently look into the demographic data on which students were dismissed or did not pass courses. While there was certainly a place for this kind of dialogue, the assumptions inherent in it were unsettling. Why does a discussion about anti-racist teaching and evaluation practices immediately lend itself here? Why were the deficits of the plenty of White students who occupy a majority of Students of Color's classrooms not a part of the discussion? Why not discuss strengthening grading and evaluation for the cultural competencies we *do* expect of our students and faculty? In my experience as an educator, it was disheartening to see how few expectations we had of incoming students in terms of their cultural awareness and sensitivities. Why were we not working to strengthen our expectations of anti-racist training, thereby improving our students' experiences?

Some of these questions were, unfortunately, answered. Not directly, but by a colleague's admission that she "doesn't know what she doesn't know" as she pleaded for the institution to offer her as faculty more opportunities for DEI training. On the one hand, this vulnerability expressed by a person in a position of significant power was important. On the other, I wondered if she would feel as comfortable, as a tenured faculty member, asking for training on other basic requirements of her job in front of all of her peers and leadership? If we focused on increasing our expectations of our students, our own cultural incompetence might be exposed. Thus, it may be safest instead to mitigate issues of bias around student deficiencies.

One of the unfortunate results of this thinking is that it furthers the minority tax on colleagues from already marginalized identities. If I, as a White counselor educator, am approaching my students of color with fear of challenging them as the human student beings that they are (who pay considerable money for me to do just that) out of fear of being perceived as perpetuating negative implicit biases, that often means leaving their most meaningful training needs to my Colleagues of Color to manage. The literature is clear that my Colleagues of Color will receive less institutional support with this increased load of remediating trainees, often leading to their own burnout (Hirshfield & Joseph, 2012). One can see how this cycle perpetuates itself by leaving the Faculty of Color to take on any major challenges presented by Students of Color, rather than meaningfully sharing this load through developing deeply trusting and challenging relationships in a collective community.

The Difference of Synergy: Restoration

Education, especially in healing, should teach students how to reach within, amplify, and hone the healing resources inherent and instilled in them (hooks, 1994; Wall Kimmerer, 2015). How are we currently doing that? When DEI training is devoid of love, soul, and creativity, we perpetuate the very harms of cultural supremacy (Engram & Mayer, 2023). We increase trainees' vocabularies, teach them to critique, and to be critically conscious. But like cultural supremacy itself, when we do not engage students' sense of humanity, freedom to act creatively, and help them experience the joy of deep, complex human connection, we train counselors who believe simply integrating "objective" healing approaches with critical consciousness is enough.

In contrast, Indigenous approaches to healing are, at their core, concerned with *restoration* and harmony (Katz, 2017). Malcolm X is famously quoted as saying, "If you stick a knife in my back nine inches and pull it out six inches, there's no progress. If you pull it all the way out, that's not progress. The progress is healing the wound that the blow made… And they won't even admit the knife is there." If something is not working, restoration means healing and creating anew. I imagine it was not comfortable for Katz and his colleagues to communicate to their university that *the* Paolo Freire would not actually offer them a lecture and instead would participate in small groups as any other member of campus. If our mental health care system is not working, true restoration would mean that its future looks different. In order to release untapped healing energy and create something new, many of us have to undergo some grieving in order to release the ways of being and knowing that we have previously held onto.

My Comfort Zone

My grieving process began shortly after I graduated from my own doctoral program in Counselor Education & Supervision and began working full-time under Dr. Gillian Berry at a college counseling center. I felt a general sense of confidence, having had difficult cross-cultural dialogues around racism, and having accepted and integrated feedback about my impact on others as a White person. However, I had no clue how deeply entrenched my entire schooling and clinical training—even in regard to topics of DEI—were rooted in one dominant way of knowing. Even with what I thought then was a critical eye for social justice, I had no sense of the baggage and assumptions I was carrying from past work experiences until I was forced to face them directly.

I was a recently licensed therapist, weeks into my new role at the counseling center, when I was first introduced to the SC2.0 model. I was transferred a call with a concerned mother who asked to speak with me. Probably flattered, I listened to the mother's concerns and encouraged her to have her child come to the center at the very start of my next walk-in shift. The mother thanked me and let me know that she chose to ask for me specifically after viewing our counseling website; "you look just like my daughter," she said, "I just feel like she'll connect with you best."

It was not until I went to document the call that I saw that the mother had been referred to my colleague, Dr. Summer, by a trusted colleague in another department. Summer was about ten years more experienced than me, with credentials I did not have in the specific area her child was struggling with, all of which was listed on our website. Summer was the appropriate referral. I soon realized that after the mother had spoken to Summer, she looked at our website, saw my face, and decided to call and ask for me instead. I understood what the mother had meant by her comment about "just knowing" I would be the right fit for her daughter, having never met me or my colleague, from our faces in the photos on our website: mine White, "just like" her daughter, and my colleague's, Black.

Having not yet become familiar with the organizational culture of the counseling center, I felt uncomfortable simply moving forward with the client, knowing now that my colleague was expecting to meet with the same student. I decided to share what had just transpired with Summer, who tilted her head with a knowing look as soon as I said, "looks just like you." She thanked me for being up-front with her about what I was told, shared her unfortunately unsurprised reaction with me, and let me know she was more than fine with me working with this particular client. She actually asked me, "How are you doing with this?" I left that conversation feeling the heaviness of how unsurprised my colleague was, yet closer to my esteemed colleague and having felt like I did as close to "the right thing" as possible. Looking back, selfishly, I probably wanted to make sure my colleague knew that I was aware of the racist undertones of that call and that she knew I wanted to move forward as a team. I looked up to this person.

The walk-in shift that I was expecting the student to come to came around later that week. After waiting about 20 minutes of grace time, the student still had not shown up. Other students were waiting, so I informed our front desk that if the student I was expecting did walk in, they should please let her know that I would be with her once I was done with a student in crisis.

After tending to the crisis, I saw that the student I was expecting had arrived—and was visibly frustrated. Having just concluded training in the SC2.0 model, I met her frustration with a calm "Thank you so much for your wait, I'm glad you came today." The student let me know she had been "made to wait for over an hour." I briefly explained the nature of walk-in hours and let her know that when she still had not arrived 20 minutes after I informed her mother that my walk-in shift started, I needed to see students who had been waiting. The student insisted that she arrived on time, to which I simply said, "Okay, I am not sure what happened then, but you're here now. Let's talk about what's going on for you."

It did not take long for the student's demeanor to shift, and for us to get some work done. As we ended our consultation, the student stood up, thanked me, and offered a reassurance: "I'm sorry I was so frustrated when I came in. I know you didn't do anything wrong." I reiterated that I was glad she had come in, and that I would look forward to seeing her again if and when she needed.

Not a minute went by and my colleague, AJ, knocked on my office door. He had been in his position at the center for nearly 40 years, checking in hundreds of students every day. He was shaken and visibly upset. "We need to talk," he said sternly. He shared that the student who had just left my office had walked back to the front desk and berated him in front of the entire waiting room before leaving, adding that "even my counselor said it was your fault!" This colleague, like the one the client's mother preferred me over, was Black.

I knew enough to avoid being defensive—why would he think I said any differently after being so embarrassed in front of the entire waiting room? He did not know me. He had no context about me other than I was a new, young, White colleague. "That is awful, and I am so sorry," I said. "I do want you to know that I would not say that to a student." I shared with him what I had communicated to the student, and acknowledged that my saying, "I'm not sure what happened, then," may have opened the door for the client to not see us as a united team.

I may not have told the student what she said I did to AJ, but my neutrality and desire to just move past it may have given her the impression that if *I* did not do anything wrong, then certainly *someone* had—and she could feel entitled to say so. The possibility that she may have delayed her own appointment was inconceivable to her (a timestamp that recorded her arrival did, in fact, record her as arriving 25 minutes late), but had I felt more confident in the SC2.0 model, I may have mitigated the issue by giving more clear instructions to her mother about how our process regarding late arrivals worked. I could have chosen not to honor the family's preference to see me and let them know that if the student arrived at the start of my walk-in shift, I would see her, but otherwise, I would take the next available student, and there would be plenty of available clinicians who could help her out. AJ continued to share how upset he was. My explanation did not seem to matter much to him at that moment, and I understood that trying to justify myself was not the point. I settled into just listening as best I could.

Up until this point in my life, I had learned enough about my social positionality to manage my "White fragility" (DiAngelo, 2018) in these situations. I knew the difference between intent and impact and how little the former mattered. I knew how to receive feedback. The situation was sad, painful, and uncomfortable, but it was not new to me. What happened next, however, was.

The Difference

A few minutes later, I heard a conversation grow increasingly heated between AJ and my new counseling center director, Dr. Gillian Berry. Remember, I had only begun working there a few weeks before. Gillian seemed to increase her volume with his volume, at times interrupting AJ, and it felt to me like she was exerting her power and position of authority over him. My first thought was, "Why is she confronting AJ about this? He just got humiliated in front of a student. If anything, I should be the one she's confronting right now."

As I stood in the hallway, dumbfounded, my colleague peeked out her office door and asked me what was going on. "Honestly, I don't know," I said. "Gillian is going at it with AJ, but he was just yelled at by a student who left my office." I told her what the student's mother had said to me on the phone, their consult with Dr. Summer, and how the student responded to me in comparison to AJ. "If anything, it's my fault," I said.

"Oh, that's horrible," my colleague said, also with a knowing look.

"What's up with Gillian? Does she do this often?" I asked. "Should I interrupt them and say something?"

My colleague shrugged calmly. "Oh, yeah, I wouldn't worry. She's got it."

Another colleague joined us. She responded in almost the exact same way.

I was perplexed. Was everyone in this office okay with AJ getting the brunt of this situation? Or was something else happening that I could not see?

I returned to my office, and shortly thereafter, Gillian knocked on my door. She sat down, looking tired, and let out a deep breath.

"How are you, sweetie?" She managed to say.

I should say at this point that all of these interactions were already in contrast to the director I had worked for only a month previously, who missed no opportunity to remind her staff of her authority.

I immediately dove into what happened, I am sure speeding through my words. I began with how AJ had done nothing wrong, and how I was open to feedback on how I could have shown up differently with the student to have prevented that interaction in the first place. Gillian nodded calmly and said, "Okay." With a kind smile, she left my office.

"Where am I?" I thought to myself.

Grieving the Comfort Zone

I expected to be managed, reprimanded, and educated on how my social positionality played into the issue at hand. Instead, I was spoken to calmly and kindly, was called sweetie, and told that it was "okay." I could think of no other work environment where this would have been the response from my director in such a situation. On the one hand, I could see some directors being so oblivious to the racial dynamics that they would not be someone that AJ would speak to at all, and on the other hand, I could imagine directors addressing this issue by trying to validate and soothe AJ and confront me with a traditional DEI conversation about implicit bias and culpability in institutional racism. I was so convinced that the latter was the right and only way that the possibility of any other approach having merit was beyond my comprehension. In my mind, the only explanation was that Gillian, despite being a Black woman herself, just berated my Black colleague who had been in his position almost as long as I was alive, and let me, her new young White employee, completely evade responsibility. If this was not a case of upholding the status quo of a racist institution and internalized racism, I did not know what was. What gave me enormous pause, however, was the three colleagues in a row, Dr. Summer included, whose reactions to the interaction between Gillian and AJ were calm, secure, trusting, and knowing; "Gillian's got this."

Taking Responsibility

I later learned that Gillian's heated conversation with AJ went something like this: AJ approached Gillian upset, shaken, and shared what had happened with my student. (This is notable, as Gillian was not even who AJ formally reported to at the center. So, his approaching her from the outset was from a place of comfort and need for support, not expectation or because of her technical position of authority). Gillian spoke to AJ with a fierceness to remind him of who he was, what he brings to the center every day, and how she was not going to let that student's berating distract him from knowing his absolute, unquestionable worth. She cut him off when his worrying began to spiral, she raised her voice over his to take charge of the conversation, knowing he was seeking her mentorship. Gillian was not lacking in awareness about the context of racism that orchestrated the entire situation. Instead, she allowed her conviction to fuel her to help AJ remember who he was. To uplift him.

By the time Gillian got to me, I got an "okay." I later realized what felt so off to me about her response to the entire thing. It was that *I was not centered.* My needs as a White person were not subtly prioritized. Her energy was not focused on me, it was focused on where it was most needed for healing.

By the time Gillian got to me, instead of accepting her offer to share how I actually was doing, in response to "How are you, sweetie?" I told her everything I *already* knew. I completely blocked her from accessing *me*, with all of the analysis and terminology I had about the racist interaction, to show her I understood what happened and to try to stop her from intervening further with AJ.

Looking back now, I wish I could have been vulnerable. I wish I had simply said, "Gillian, this all happened so fast. I really like working here so far, but I still feel really nervous and new. I felt this whole racist interaction coming from earlier this week, and I see how easily I was used as a pawn in it when all I wanted to be able to do was interrupt it. I feel horrible. I feel helpless. I hate how quickly this all happens, and I wish I had the agency and skills to use my positionality more effectively. But I don't, and look what happened. I feel awful." Knowing the Gillian that I know today, this would have led to an unforgettably powerful interaction. But I was not ready for one. And she saw that. And if there is anything I know about Dr. Gillian Berry, it is that she is not in the business of forcing someone to be ready for what she has to offer. She knows the dosage of healing she offers requires consent. But she also knew something else in that setting, and it was the felt sense of *collective responsibility* she had cultivated at the center among her staff. Gillian left that conversation knowing I had some growing to do, but did not feel the need to be the person to do it.

Not a week later, one of my colleagues, our diversity coordinator, ran a prescheduled in-service for our team. This in-service was voluntary, led by my colleague and our diversity committee, with support from Gillian only if needed. As a member of the diversity committee myself, I also had some say in what we could do in the in-service. Even as a new employee and White staff member whose White student just berated our Black front desk staff (which by that point was a well-known interaction amongst the center staff), who clearly had so much to learn, I was welcome in that committee, and my ideas were encouraged and respected. Gillian participated in these sessions without knowing the prompts, just like everyone else on staff, trusting her employees to facilitate and lead with their own strengths. We did an activity our diversity coordinator introduced, which focused on social positionality. This activity and ensuing dialogue allowed my colleague Dr. Summer and I to share what had recently happened with the student's mother and the student.

Summer shared how the situation frustrated her; after all her years of training and expertise: Just by looking at her face, the client's mother decided she was not the right fit. My Colleagues of Color shared their experiences dealing with the same in and out of the counseling center. People shared similar experiences across marginalized identities. We discussed how, as a racially diverse team, we could work better together to reduce the burden our colleagues of color experience and also take steps to interrupt the institutional racism that we lived, worked, and breathed in every day. The floor was open, the discussion deep and meaningful, and fully run by staff members. Gillian chimed in with her personal experiences, but she need not lead the discussion. She trusted her staff to work through these things together. And we did. And we were able to be open with our limitations and needs with one another. We were able to have healing dialogues that made it possible for us to help heal our student body, existing in the same sociopolitical turmoil as we were.

She did not need to do anything more than an "okay" at that moment with me. She knew that the work she put into the culture meant an environment was created such that I could not actually escape this learning if I tried. And that more learning was available, the more I was ready to take responsibility for my own growth. And the collective staff shared that responsibility. This regenerative healing resource is synergy in action.

Collective Healing and Restoration

Years later, when reflecting on this experience with Gillian, I recalled that my own Sicilian–American grandmother actually assumed a presence like that of Gillian in her confrontation with AJ in how my grandmother showed us love. She had an intensity to her, a power and an impact. That was what I was accustomed to love looking like. Yet, somehow, that was trained out of me in my formal schooling and clinical training in favor of a more passive, Rogerian clinical presence in which one actively avoids exerting power (Hill & Lent, 2006). Despite the growing diversity of researchers on therapy, therapists, and therapists in Western professional helping training since Rogers' time, helping skills training programs today remain grounded in his approach. Hill and Lent (2006):

> The helping skills training programs developed more recently (e.g., Brammer, 1973; Brammer & MacDonald, 2003; Cormier & Cormier, 1998; Danish & Hauer, 1973; Egan, 1975, 1994; Evans, Hearn, Uhlemann, & Ivey, 1998; Gazda, Asbury, Balzer, Childers, & Walters, 1977; Goodman & Dooley, 1976; Hill, 2004; Hill & O'Brien, 1999; Meier & Davis, 2001; Okun, 2002) were all variations of training programs that were developed in the late 1960s and 1970s (e.g., Carkhuff, 1969; Ivey, 1971). These early training programs were all based loosely on the foundation of Rogers' (1942, 1951, 1957) client-centered therapy, which focused on therapist facilitative conditions as being the necessary and sufficient conditions for effective therapy. Although Rogers (1942) first considered the facilitative conditions as skills (e.g., restatements and reflections of feelings) that could be taught, he later came to the conclusion that they were attitudes that could not be taught (Rogers, 1957). Many of his followers (e.g., Carkhuff, 1969; Truax & Carkhuff, 1967), however, believed that the facilitative attitudes were skills that could be taught and hence developed systematic training programs to teach them (p. 155).

In fact, Parham and colleagues (2016) warn that therapists so trained may be ill-equipped to work with culturally different clients who might find a more active approach relevant to their own needs and values. In a study conducted by Berman (1979), it was concluded that "Black and White counselors appear to adhere to two distinctive styles of counseling." When shown a videotape of culturally varied clinical vignettes and then asked, "What would you say to this person?" at the same point of the session, White trainees recommended more passive attending skills significantly more than Black trainees, who tended to recommend more active expressive skills. Thus, trainees who are not from the dominant culture Rogers' style and approach was borne from (White, male, cisgender, native English speaking, US citizen, etc.) may be trained out of their authentic counseling style in favor of the dominant passive attending skills that continue to be emphasized in counselor training. While this study was conducted in 1979, I see the same authenticity being trained out of my students in favor of alignment with the more dominant approach. Our current training programs are still grounded in the microcounseling skills developed in the 1960s and 1970s (Hill & Lent, 2006).

If I am honest, though, I was trained to be embarrassed by my grandmother's unique presence in my sociocultural environment much earlier than I was exposed to counseling training. The way she walked down our elementary school hallway with her high heels clicking, shoulder pads, colorful outfits, and dressed like she was going to a fine dinner rather than an elementary school office. The way she permed her hair every two weeks and kept it jet black. The way she terrified my friends when she got on the elementary school loudspeaker in the middle of the school day and frankly said, "To the boy who bit my granddaughter, report to the principal's office immediately" when I was the one who accidentally hit a boy in his mouth during a recess soccer game and hurt my hand. Or when another school aide walked me to her office after seeing me eat a potato chip off the cafeteria floor, and my grandmother simply looked at me incredulously and said, "Mama makes you homemade pasta. Why would you eat a potato chip off the dirty floor?" I loved her and was in awe of her, but I never really knew what was about to come out of her mouth. And at times, I am embarrassed to admit now, I wished she acted a bit more like the other grandmothers I was surrounded by, who were more soft-spoken, who let their hair turn gray, and who my friends were not intimidated by. This is such a minor example in comparison to the internalized shame we know is imposed upon groups of people who do not align with societal ideals of behavior, size, language, skin tone, and self-expression. Cultural domination, or the messaging that there is one right way to be and there are consequences when showing up otherwise, certainly precedes the mental health profession. However, the gaslighting occurs when a profession is so deeply entrenched in one "right" way of being, while claiming it is possible to genuinely integrate diversity, equity, and inclusion principles in this context.

Today, whenever I smile knowingly with another colleague as Gillian gets into her unique mode of healing, I also think to myself, "She's got this." And I restore another piece of that embarrassment I held in witnessing my grandmother be the fullest version of herself, even if that was not the dominant society's prescription in my environment. Going to Gillian for support and feedback means bracing yourself for the impact of someone really seeing you and believing in you to see yourself, especially if you want to help others heal. This approach has fostered an inescapable sense of individual responsibility in those who seek her feedback. I have learned to become more and more myself in my teaching and counseling, in navigating challenges in the context of cultural domination, and more in tune with myself when I notice incongruence. When it comes to DEI, our ability to speak to "those terms and theories," as our SCS colleague said, is irrelevant. Who we are and *how* we show up as human beings in the moments that matter does instead. How we reach within and tap into our own unique healing energy, helps to release that of others. This is synergy.

Raising Our Expectations: Synergy

As illustrated in my own experience, the difference of *synergy*, or an approach to DEI that is philosophically grounded in Indigenous ways of knowing, is *felt*. Initially, when viewed from a Western philosophical lens that we have been trained to accept as not a lens at all and instead "objective," synergy in practice can appear to be color-blind or idealistic. When I was confused about why Gillian, as director of our counseling center, was taking the approach that she was with AJ, I viewed her actions through an individualistic lens, and was unsatisfied. Synergy is a healing energy that is co-constructed in the environment through collective responsibility. What my narrow perspective failed to allow me to see at the time was the sea of collective responsibility that surrounded the counseling center staff and upheld the strands of empathy, truth and reconciliation, flexibility, creativity, interconnectedness, personal agency, contextualization, growth orientation, and diplomatic disruption in the center's day-to-day operations. Through years of training, personal experience, and mentorship from this standpoint of collective responsibility, I believe that we can expand the ways we currently train and teach DEI as integrated with well-being, by helping students and trainees to better reach within and expand healing potential around them. It is my hope that the stories in this book help speak to the difference of synergy, and that we all feel a bit more empowered to bring more of ourselves into this work to release the untapped healing powers readily available in our classrooms, clinics, workplaces, and communities.

Reflection

Peter Cornish

Cultural gaslighting is most powerfully represented in the mental health crisis narrative that dominates news headlines and discourse in professional mental health settings. But the truth is there is no mental health crisis: Instead, there is a cultural and economic crisis. The mental health crisis narrative is a convenient bipartisan diversion tactic appealing to highly educated liberals as much as right-wing populist extremists. By endorsing the narrative, powerful entities and organizations across the political spectrum, invested in maintaining status quo power structures, are diverting responsibility for climate and social injustice to overburdened, disempowered, and disenfranchised consumers addicted to social media tripe and rampant consumerism. The result is a crisis of disconnection from each other, our natural environment, any form of spirituality, or a sense of being part of something larger than ourselves. There is no climate emergency. There is no systemic racism. Classism does not exist. You are not getting poorer. You are not being swindled by narcissistic right-wing billionaires. You are simply crazy. If you live in the United States, you have been getting crazier and crazier over the past two decades, as evidenced by the climbing suicide rates. Opioid crisis? No self-control. You need some serious mental health treatment—and by the way, it will be impossible to get it unless you are wealthy. A dystopian remake of the 1944 George Cukor thriller, *Gaslight*, could be overdue. Who might we cast as the villain this time?

But a dystopian thriller is not what we need. Instead, we need to open ourselves to the invitation synergy offers. Personally, I see it most clearly in nature. Not long ago, an Indigenous Elder said something at a psychology conference that changed my life. I arrived a few minutes late and was greeted by a wide circle of chairs in a double large conference room. A White man was delivering an Indigenous land acknowledgment. This was to be followed by a prayer from a traditional Elder. The Elder, a woman possibly in her late fifties, wore a few articles of traditional clothing, along with more typical conference attire. Her eyes bright, she entranced those in the room with her warmth, spirit, wisdom, and groundedness. She began, "You know, the land acknowledgment is very important. The land is of the creator and a gift to all of us. You might know better than me that a lot of people feel lonely. Sometimes, it can be worse when you travel to a new place, as all of you have done today. I never feel lonely, no matter where I go. That's because wherever I go, the trees, lakes, and rivers are there to greet me. The rocks and the clouds and the sky are there, too. These, in truth, are the same trees, rivers, and rocks that I know. They are all of Mother Earth. They are my friends. My people don't rely *only* on people to feel connected. We never feel lonely because we have so many friends."

Two things happened to me shortly after receiving this gift of wisdom. First, I realized that the northern California city I was living in was *not* the alien urban wasteland I thought it was, blighted by inequality, racism, addiction, and homelessness. No, because the same beautiful trees, rivers, and skies were there with me, too. They are the friends I know and love. They were inviting me to heal, but I had not really seen them. I had been hiking through towering groves of redwoods and riding my bicycle along shaded canyon parkways for three years by then, but I was not really *seeing* my old friends. I was not opening my heart to them. Until, that is, I accepted an invitation from a wise Indigenous healer.

The second thing that happened was that I visited my family cabin on an island on Lake Huron. The island sits in the traditional and treaty territory of the Shawanaga First Nation, which belongs to the larger Anishinabek Nation that stretches across much of Ontario, east along the Prairies, and along the northern shores of all the Great Lakes. The Anishinabek Nation is composed of peoples of the Algonquin, Saulteaux, Anishinabe (Ojibway), Odawa, and Chippewa nations, as well as others. I was preparing to welcome colleagues from the University of California, Berkeley, and Stepped Care Solutions, who would join me in a few months' time to connect with my mentor, Dick Katz. We would be filming a documentary on the island about his life and work on synergy, and how it has become central to my work and the work of my colleagues. We would also be honoring the sacredness of the land with an Indigenous prayer and a traditional talking circle. But that was still two months in the future.

I have visited this rocky island every year of my life except for two years when the pandemic restricted travel. While I have an enduring love for the place, over the past 20 years, it had increasingly become a place of grief, strife, and sadness for me. This time was different. This time, I recognized the ancient billion-year-old rocks as my friends, and the windswept pines were waving their greetings to me. The lapping water was murmuring, "Welcome back." I still felt sadness and grief, but the land comforted me, and held me through those feelings. The lamps, fueled by kerosene, would not need to be lit. I would be sleeping right after sunset.

References

American Psychological Association. (2015). Gaslight. In *APA dictionary of psychology*. Retrieved June 1, 2024, from https://dictionary.apa.org/gaslight

Berman, J. (1979). Counseling skills used by black and white male and female counselors. *Journal of Counseling Psychology, 26*(1), 81–84. https://doi.org/10.1037/0022-0167.26.1.81

Bombay, A., Matheson, K., & Anisman, H. (2014). The intergenerational effects of Indian residential schools: Implications for the concept of historical trauma. *Transcultural Psychiatry, 51*(3), 320–338. https://doi.org/10.1177/1363461513503380

Branco, S., & Jones, C. (2021). Supporting black, indigenous, and people of color counselors: Considerations for counselor skills training and practice. *Journal of Mental Health Counseling, 43*(4), 281–300. https://doi.org/10.17744/mehc.43.4.01

Brear, P., & Dorrian, J. (2010). Gatekeeping or gate slippage? A national survey of counseling educators in Australian undergraduate and postgraduate academic training programs. *Training and Education in Professional Psychology, 4*(4), 264–273. https://doi.org/10.1037/a0020714

Brubaker, M. D., Puig, A., Reese, R. F., & Young, J. (2010). Integrating social justice into counseling theories pedagogy: A case example. *Counselor Education and Supervision, 50*, 88–102. https://doi.org/10.1002/j.1556-6978.2010.tb00111.x

Celinska, D., & Swazo, R. (2016). Multicultural curriculum designs in counselor education programs: Enhancing counselors-in-training openness to diversity. *Journal of Counselor Preparation and Supervision, 8*(3). https://doi.org/10.7729/83.1124

Constantine, M. G., Smith, L., Redington, R. M., & Owens, D. (2008). Racial microaggressions against black counseling and counseling psychology faculty: A central challenge in the multicultural counseling movement. *Journal of Counseling and Development, 86*, 348–355. https://doi.org/10.1002/j.1556-6678.2008.tb00519.x

Daniel, C. (2008). From liberal pluralism to critical multiculturalism: The need for a paradigm shift in multicultural social work practice in the United States. *Journal of Progressive Human Services, 19*(1), 19–38.

Daniel, C. (2011). Lessons learned: Pedagogical tensions and struggles with instruction on multiculturalism in social work education programs. *Social Work Education, 30*(3), 250–265. https://doi.org/10.1080/02615471003789829

Deepak, A. C., Rountree, M. A., & Scott, J. (2015). Delivering diversity and social justice in social work education: The power of context. *Journal of Progressive Human Services, 26*(2), 107–125. https://doi.org/10.1080/10428232.2015.1017909

Dollarhide, C. T., Smith, A. T., & Lemberger, M. E. (2007). Counseling made transparent: Pedagogy for a counseling theories course. *Counselor Education and Supervision, 46*, 242–253. https://doi.org/10.1002/j.1556-6978.2007.tb00029.x

Ekpe, L., & Toutant, S. (2022). Moving beyond performative allyship: A conceptual framework for anti-racist co-conspirators. In *Developing anti-racist practices in the helping professions: Inclusive theory, pedagogy, and application* (pp. 67–91). Springer.

Engram, D. V., & Mayer, K. (2023). Ya'll don't hate white supremacy enough for me: How performative DEI prevents anti-racism and accountability in higher education. *The Vermont Connection, 44*(1) Retrieved from https://scholarworks.uvm.edu/tvc/vol44/iss1/10

Freire, P. (1972). *Pedagogy of the oppressed*. Penguin Books.

Goodman, D. J. (2011). *Promoting diversity and social justice: Educating people from privileged groups*. Routledge.

Goodman, R. D., Williams, J. M., Chung, R. C. Y., Talleyrand, R. M., Douglass, A. M., McMahon, H. G., & Bemak, F. (2015). Decolonizing traditional pedagogies and practices in counseling and psychology education: A move towards social justice and action. In *Decolonizing "multicultural" counseling through social justice* (pp. 147–164). Springer. https://doi.org/10.1007/978-1-4939-1283-4_11

Guenette, F., & Marshall, E. A. (2008). Indigenizing counselor education: Implementing post-secondary curriculum change. *The Canadian Journal of Native Education, 31*(1), 107–122. https://doi.org/10.14288/cjne.v31i1.196432

Hill, C. E., & Lent, R. W. (2006). A narrative and meta-analytic review of helping skills training: Time to revive a dormant area of inquiry. *Psychotherapy: Theory, Research, Practice, Training, 43*(2), 154–172. https://doi.org/10.1037/0033-3204.43.2.154

Hirshfield, L. E., & Joseph, T. D. (2012). 'We need a woman, we need a black woman': Gender, race, and identity taxation in the academy. *Gender and Education, 24*(2), 213–227. https://doi.org/10.1080/09540253.2011.606208

Jacobs, S. C., Huprich, S. K., Grus, C. L., Cage, E. A., Elman, N. S., Forrest, L., Schwartz-Mette, R., Shen-Miller, D. S., Van Sickle, K. S., & Kaslow, N. J. (2011). Trainees with professional competency problems: Preparing trainers for difficult but necessary conversations. *Training and Education in Professional Psychology, 5*, 175–184. https://doi.org/10.1037/a0024656

Katz, R. (2017). *Indigenous healing psychology: Honoring the wisdom of the first peoples*. Healing Arts Press.

Kimmerer, R. W. (2015). *Braiding sweetgrass*. Milkweed Editions.

Leahy, M. J., Rak, E., & Zanskas, S. A. (2009). A brief history of counseling and specialty areas of practice. In *The professional counselor's desk reference* (pp. 3–13). Springer.

Michaelson, V. (2021). Developing a definition of spiritual health for Canadian young people: A qualitative study. *International Journal of Children's Spirituality, 26*(1–2), 67–85. https://doi.org/10.1080/1364436X.2020.1856048

Mokgobi, M. G. (2014). Understanding traditional African healing. *African Journal for Physical Health Education, Recreation and Dance, 20*(Suppl 2), 24–34.

Moodley, R., Sutherland, P., & Oulanova, O. (2008). Traditional healing, the body and mind in psychotherapy. *Counselling Psychology Quarterly, 21*(2), 153–165.

Paluck, E. L., & Green, D. P. (2009). Prejudice reduction: What works? A review and assessment of research and practice. *Annual Review of Psychology, 60*, 339–367. https://doi.org/10.1146/annurev.psych.60.110707.163607

Singh, A. A., Nassar, S. C., Arredondo, P., & Toporek, R. (2020). The past guides the future: Implementing the multicultural and social justice counseling competencies. *Journal of Counseling & Development, 98*, 238–252. https://doi.org/10.1002/jcad.12319

Smith, S. D., & Ng, K. M. (2009). International counseling trainees' experiences and perceptions of their multicultural counseling training in the United States: A mixed method inquiry. *International Journal for the Advancement of Counselling, 31*, 271–285. https://doi.org/10.1007/s10447-009-9083-3

Smith, T. B., & Trimble, J. E. (2016). *Foundations of multicultural psychology: Research to inform effective practice*. American Psychological Association. https://doi.org/10.1037/14733-000

Sowbel, L. R. (2012). Gatekeeping: Why shouldn't we be ambivalent? *Journal of Social Work Education, 48*(1), 27–44. https://doi.org/10.5175/JSWE.2012.201000027

Walker, M. (2008). How therapy helps when the culture hurts. *Women & Therapy, 31*(2-4), 87–105.

Williams, M., Dunlap, M., & McCandies, T. (1999). Keepin' it real: Three black woman educators discuss how we deal with student resistance to multicultural inclusion in the curriculum. *Transformation: A Resource for Curriculum Transformation and Scholarship, 10*(2), 11–23.

Young, G., & Davis-Russell. (2003). Dealing with difficult classroom dialogue. In P. Bronstein & K. Quina (Eds.), *Teaching gender and multicultural awareness: Resources for the psychology classroom* (pp. 347–360). American Psychological Association.

Chapter 5
A Journey of Discovery

André Forget

Part 1: Synergy

In the conference room of a Holiday Inn on the edge of Parry Sound, Ontario, Dr. Richard Katz stands in a shaft of October sunlight pouring in through the plate glass windows. He is about to give a lecture about synergy. The academics and mental health workers seated in the two rows of chairs in front of him have also traveled far—some all the way from the University of California, Berkeley—but Katz does not start by talking to them. Instead, he turns to the film crew, who are standing at the edge of the room, shooting footage of the lecture for a documentary.

"Tell us who you are and what you're doing," he says encouragingly to a burly man with a video camera strapped to his chest.

"My name's Byron," the man answers. "I'm from Halifax, but I've been living in Toronto for the past seven years. I'm a camera operator and director of photography."

I can't see what this looks like from Byron's camera, but I'm sitting behind the other camera operator, an Australian named Alice, so I can see Katz's miniature double moving around on the monitor of her rig. We aren't even five minutes into the lecture, and Katz has already broken the fourth wall.

"I have a particular thing with these people here called the crew," Katz explains to the audience, "because they're part of the picture." The film crew may be here to record the proceedings, but they're not going to do so as proverbial flies on the wall. Neither, as a writer, will I. He moves on to the next person in the crew, a tall, bearded sound engineer who says his name is Adam. This gives Katz pause.

"I have to just say one quick thing. That's the same name as my deceased son—so I have a very special connection to that man, not because of who he is, but what his name means."

A. Forget (✉)
Sheffield, South Yorkshire, UK

P. Cornish, G. Berry (eds.), *Synergy*,
https://doi.org/10.1007/978-3-031-88493-1_5

Katz continues around the circle, giving everyone a chance to say who they are, where they're from, and what has brought them here. He speaks gently, interjecting when he has something to say and asking questions when they occur to him. He is a slight man, with a neat trapezoidal mustache dominating his upper lip. His eyes, which are set deep in a tanned and craggy face, are bright blue, and his voice betrays the slight trace of his long-distant Brooklyn boyhood. It is nearly impossible to believe he is 85 years old. But having read his most recent book, *Indigenous Healing Psychology*, I know that his long and remarkable career began at Harvard in the early 1960s. He has observed many of the key developments in post-World War II psychology from the front row, and he was a direct participant in at least a few of them. But unlike his peers who slid comfortably into Ivy League sinecures, Katz—who prefers to go by Dick—followed a restless path of his own choosing, a path that has taken him from the Kalahari Desert to Fiji, Alaska, and northern Saskatchewan, from the mainstream psychological frameworks of Maslow and Erickson to far older traditions practiced by the Indigenous healers he refers to as "our First Psychologists."

The leisurely cycle of introductions is not merely a way of breaking the ice; it is part of what Katz has come here to talk about, which is how Indigenous philosophies and approaches to healing can be brought into conversation with mainstream Western psychology. Going around the circle and reminding everyone that the people recording the session (myself included) are also *participants* in the session is an aspect of the lecture itself. We are each of us factors in the creation of this particular environment in this particular place for this period of time, and together we are creating something—a mood, a sense of camaraderie—that is larger than our own individuality. This is literally what the word synergy means.

When we are finished going around the circle, Katz starts talking more explicitly about what we've just done. "When I was younger, I thought the introductions were the most boring part of any presentation," he says. "But how are we possibly going to create something larger than the parts, where what's good for one is good for all, if we don't even have an idea what the parts are? … That's the key, not to smooth over differences, but to try to understand the differences and respect them. And that's the key to synergy, to feel the lack of comfort because someone is different."

In this same spirit, Katz starts into another set of introductions—this time to people who are not physically in the room, but whose influence can be felt in every story he tells. One by one, we see the pictures of four men projected onto the wall. The first is of =Oma Djo, a traditional healer from the Ju/'hoansi in the Kalahari Desert, who Katz first met in 1968, when he traveled to the region to conduct research. The second is of Joe Eagle Elk, a Lakota healer from the Rosebud Reservation in the state of South Dakota. The third is of Ratu Civo, a healer from Fiji in Melanesia, and the fourth is of a smiling man named Danny Musqua.

"If Danny were with us, we would be traveling together," Dick says. "We always travel together. I would not give a talk drawing on Indigenous knowledge without Danny starting off."

The reason Musqua, an Elder, author, healer, teacher, and member of Keeseekoose First Nation, is not present in the room at the Holiday Inn is that he passed away the previous year. This is a loss that Dick speaks about in somewhat conflicted terms. Musqua, he tells us, had been both a "spiritual father" and a brother to him. They lived near each other in northern Saskatchewan, and worked closely together as teachers and healers. Dick's bereavement is palpable. But he also talks of his friend in a way that makes it clear that Musqua continues to be present for him—that there is a clear sense in which Musqua *is* in the room with us, because his wisdom and his spirit live on.

The physical absence of Musqua does, however, complicate to some degree the work that Dick has come to Ontario to do. Dick has spent the past 55 years trying to bring the wisdom of Indigenous psychological traditions into conversation with mainstream clinical psychology, and he is acutely aware that white, Ivy-league-trained academics like himself do not have an exemplary track record when it comes to studying Indigenous civilizations. This is one of the reasons he has chosen to begin his talk by telling the audience about the teachers who formed him. "The most important thing," he says, "is to connect to the people who taught me."

Dick takes great pains to communicate the richness of what he has learned from his Indigenous teachers, while navigating the complex cultural politics that are inevitably involved in this kind of work. There are some things, he explains, that cannot be shared with just anyone. This is perhaps one of the sharpest distinctions between the colonial, free-market approach of contemporary academia—which seeks to turn wisdom into information, and information into a product—and the Indigenous approaches to knowledge that prioritize spiritual growth and relationality.

Dick's mission (which is also the mission of Stepped Care Solutions, the company that put this event together) is not simply about encouraging clinical psychologists to accept the validity of the treatments practiced by Indigenous psychologists, but to overcome the Western mental healthcare establishment's deep cultural myopia, its blind devotion to metrics and clinical trials as the ultimate source of knowledge. Healing, Dick believes, is a spiritual matter. As long as spirituality is bracketed out of the healing process, people won't fully heal. And though Dick is loath to provide a narrow, textbook definition of spirituality, it is clear that he sees it as being a matter of connection—between individual humans, and between humans and the non-human creatures with whom we share our environment.

Dick's ideas about synergy flow out of this primary belief in the importance of this spiritual dimension. "One of the things that happens in the Kalahari," he says, after explaining the Ju/'hoansi practice of channeling n/om, or healing energy, "is that when I work to, let's say, heal you, you then become able to heal someone else—or the energy that goes to you gets expressed. So, the more that I heal, the more healing there is, as opposed to the model where the client comes in, you close the door, and once that door is closed, the more I work with that client, the less anyone else has access to me." Synergy, Dick stresses, is about multiplication rather than division, abundance rather than scarcity.

As I listen to Dick speak, I think back to the conversations I have had with him over the past couple of days. I think about how I read his book alone in my apartment, trying to figure out what he was driving at. I must admit that the language of spirituality is one I have a complicated relationship with. I am generally much more comfortable with material causes and material effects. But in the time I have spent getting to know Dick, I have relaxed a bit. He has traveled a long and difficult road to arrive at his conclusions, and if his road has been different from mine—just as it is different from everyone else's in the room, just as all our roads are different—this does not mean I cannot learn from it.

Dick's talk goes in several directions. He touches on structural problems, on the question of what counts as evidence, and on the ways that different Indigenous nations think about healing. Like wise men going back thousands of years, his preferred method of making a point is to tell a story. Whether he is talking about sweat lodges or talking circles, Dick is always also talking about the meaning of synergy.

I was in Parry Sound because I'd been asked to write about synergy. And spending a week with Dick convinced me that the best way to answer the question "What is synergy?" is to tell a story: the story of what it was like to be in northern Ontario with Dick, the film crew, and my colleagues at Stepped Care Solutions. At a basic level, it is a story about discovering the latent potential for positive interaction between different beings, and letting that potential be realized. This is not easy to achieve, because most of us expend a lot of energy in building up and maintaining defenses against the dangerous world outside, and while these defenses might keep us safe, they also keep us isolated. Synergy is powerful because it takes all of that energy we have been storing up for ourselves, and releases it for use by others.

This sounds very nice. But as my colleague Dr. Gillian Berry, one of the vice presidents of Stepped Care Solutions, asks later on, there is still the key question: "Can it work?"

Part 2: The Road North

When I'd flown from London to Toronto three days earlier, I had not quite known what to expect from the assignment. I'm a Canadian by birth, but I live on the edge of the Pennine Mountains in northern England, and traveling back home always seems to involve a great deal of messing about with canceled trains, delayed flights, and irritable border guards. To make matters worse, I was at the tail end of a nasty head cold. When I woke in my hotel room the morning after I arrived, I looked out over one of the sprawling parking lots that surround Pearson International Airport and felt a tremor of dread. What, I thought to myself, have I gotten myself into.

The chain of events that led to my being in that conference room with Dick had begun a few months earlier, with a call from Gillian. She asked if I might be interested in spending a week on an island in Ontario, learning about Katz's work with practitioners of Indigenous psychologies. Stepped Care Solutions would be shooting a documentary about Dick on an island that belonged to the family of Dr. Peter Cornish, the founder of SCS. Peter had recently developed a new framework for delivering mental health care that he called Stepped Care 2.0 (SC2.0), and while I had no professional background in the field of mental health care, I'd spent a year working with Gillian and Peter in a scribal capacity. I'd had my own tussles with

Canada's patchwork system of mental and primary care, and I found their critiques of the narrowly Western approach to psychology compelling.

I had started working with SCS because I needed a job, but the more time I spent helping them write about SC2.0, the more I felt that they were articulating something that I recognized, deep inside myself, was true: The crisis in mental healthcare is not simply a matter of poor resource allocation or ineffective treatments. It is, like so many of the crises we face in the twenty-first century, a crisis of Western modernity. And it cannot be addressed without asking philosophical questions about what terms like "wellness," "psyche," and "self" actually mean.

I'd spent many hours with Gillian and Peter over the previous year and a half, but I'd never met them in person. Traveling to Ontario, I thought, would be a good opportunity to get to know them better. And as I'd be writing about Dick's work anyway, I could save myself some time and effort by learning about his ideas firsthand. It didn't hurt that the island was just north of Parry Sound, and there is no place in this world quite as beautiful as the eastern shore of Georgian Bay at the beginning of autumn. I told Gillian to go ahead and book my ticket.

Parry Sound is about 200 kilometers north of Toronto, and Peter's island is another 50 kilometers north of Parry Sound. The plan was for me to meet with Dick at the airport, at which point I'd travel up with the film crew and some of the other SCS staff. I had never met any of these people before, though I'd helped one of them, Alexa Bol, with an essay she'd written for SCS the previous year, and I'd exchanged a few messages with Kareem South, who would be driving. My life in England is fairly solitary, and I spend most of my time writing and reading. I was still coughing a lot (from the cold), and coughing always makes me feel like I'm being incredibly rude. A wave of anxiety washed over me as I made my way to the gate where Dick would be arriving. A week is a long time to spend with a group of strangers.

And then a strange thing happened: I walked up to the trio of people waiting for Dick, and immediately felt the anxiety dissolve. Alexa, Kareem, and Sarah Harte gave me a welcome that was so warm, and they seemed so clearly relaxed with each other, that I found myself chatting away as if I hadn't a care in the world.

Dick came through the gate a little later. My first impression was that he must be quite tired after all this traveling. He'd been unable to get a direct flight from Saskatoon, so he'd had to travel through Ottawa. This would be an irritating amount of flying for anyone, but I imagined it must be even worse for a man of his age. But he seemed unbothered by the ordeal, and after our introductions, we made our way to the airport doors to figure out travel arrangements. The film crew had parked their big SUV by the entrance ramp, and we did another quick round of introductions with them. The leader seemed to be a petite, intellectual woman named Sara Alfaro, the director of the documentary. They would be shooting an interview with Dick on the drive north, and it seemed there wouldn't be room for Dick, the camera, and myself in the van, so I would travel with Kareem, Alexa, and Sarah to the rented cottage where they were staying not far from Peter's island. This would mean missing out on one of the big early moments of the film, Dick's reunion with Peter, who had been one of his graduate students back in the 1980s. But I'd be able to join them on the island later in the evening.

The highway from Toronto to Parry Sound cuts due north through the suburban sprawl of the GTA and over the Holland Marsh into the hill country near Lake Simcoe that marks the divide between east and west, north and south in Ontario. We talked about our lives and our families, and discussed whether or not one's placement in the family (oldest or youngest among siblings, for example) had an impact on your personality. Alexa, as the mother of five children, seemed to think it did, and I, as the second son in a family of six, was inclined to agree. After we stopped for lunch, the topic of basketball came up, and I discovered that Kareem didn't just run basketball clinics and do basketball outreach—he had played professional college ball for Berkeley. I had been struck by his spiritual sense of calm, the deliberate and thoughtful way in which he spoke, and I knew he had studied public health at Berkeley, but I began to appreciate that there must be another side to him as well, a grit and competitiveness. It's always extraordinary, when meeting new people, to discover just how many sides we all have. And I thought about how, under different circumstances—if we'd all been random passengers on a bus, for example—our two-drive north might have happened in silence, the sense of connection between us unknown, unacknowledged, and unexplored.

North of Lake Simcoe, the landscape becomes more rugged. Rolling hills and fields give way to the bare rock of the Canadian Shield and dense forests of birch, maple, pine, and cedar. The greenery of summer has turned into the glorious red and gold of autumn. In places, the highway cuts through canyons that were blasted into the rock back in the 1950s, when the road was being expanded into one of Ontario's main north-south arteries. Periodically, one sees little memorials on the side of the road, shrines erected for travelers who have died in accidents and collisions.

We pass Parry Sound and turn down a smaller road that eventually takes us to a dirt track. At the end of the track was the cottage that Alexa, Kareem, and Sarah would be staying in along with their colleague Catie Greene, who was driving up from her home in Washington, D.C. Once we settled in, we all walked down to the dock on the lake. After three days of traveling, the stillness was incredible. I felt muscles unclench that I hadn't even known I was clenching.

Part 3: Peter's Island

Nares Landing is not a town, or a village, or even a hamlet. It is a collection of buildings at the head of an inlet, with a concrete embankment, a narrow sand beach, and a jetty where boats can dock. It exists to serve the cottages dotting the islands and shores of the inlet, most of which can only be reached by water. The jetty's docking bays are marked with the names of families who own property on the inlet. The island I was heading to has been in Peter Cornish's family for more than 100 years. The autumn air was warm, but the jetty was nearly empty. Canadian Thanksgiving, the traditional end of cottage season, was only a week away. The few people still up here were getting ready to close up their properties for winter.

Night was falling, and the darkness came in from the east like a curtain being pulled across a window, but the light that remained was clear as varnish. Some of the film crew were waiting on the embankment with equipment that needed to be shipped to the island. They would be shooting in the morning, and the director, Sara Alfaro, and videographer, Alice, were staying at the cottage tonight so they could scout locations early in the morning. Their work is tied much more closely than mine to natural rhythms. They need to know about view lines and angles of sunlight. They need to be able to anticipate where to position the cameras. The contours of the earth shape how the camera operators will need to move, and the wind in the trees needs to be accounted for in the placement of the boom mic. I had never observed a film crew up close, and I was struck by the element of uncertainty present in their work. They talk about filming as if they are botanists collecting plant samples. Their job is to get as much footage as they can, and then to try to figure out what the story is once the shooting is finished.

It was getting properly dark by the time Peter's boat pulled up to the jetty, cutting through the little waves like an old silver fish. Like all boats, it has a name: the "Pretty Ugly," or P.U. I was surprised by how relaxed he looked. His gray hair hung over his forehead in a fringe of curls, and he was wearing a colorful button-up. He greeted me warmly, and we started loading the food and camera equipment onto the boat. This was a somewhat delicate procedure, as the boat kept yawing from side to side, and I hadn't been on anything larger than a canoe in years. But Peter grew up on the water, and has an intuitive sense for how to maintain his balance on a moving vessel. We pulled away from the jetty and turned west, toward the glow of civil twilight. We picked up speed as Peter opened the throttle. The bow rose smoothly out of the water, and he expertly navigated between the shoals and outcrops of rock that choke the narrows of the inlet.

The trip is not a long one, perhaps ten or fifteen minutes, but the fresh air, the lake spray, and the forms of islands and cottages looming out of the darkness gave it a sense of mystery and wonder. We were going to *the island*, a place separated from the world of train stations and airports and office blocks and fast-food restaurants not only by water, but by an invisible temporal barrier. I had spent enough time cottaging in Ontario to know that something happens to people on these small islands, even if they're only there for a couple of days. Minutes pass more slowly. Intimacies that would take months to build grow within a matter of hours. Thought is slower, too, but also clearer and more precise. This, I think, is why Peter and Gillian chose to shoot their documentary there. Synergy can be explained anywhere. But it is perhaps easier to *understand* in a place that stands at some remove from the grinding, mechanical rhythms of modern life.

As we walked up to the cottage from Peter's jetty, I was heartened to see that the buildings are more cabin than cottage. The buildings were squat and dark, with low ceilings and no electric wiring and a 1970s sort of feeling. Inside, the main cottage was lit by the warm glow of oil lamps. Dick and Gillian were sitting at the table with Sara and Alice. In person, Gillian was just as friendly, energetic, and candid as she was on our Zoom calls. I was at first surprised to learn that this was her first time being in the same room as Dick: Sitting next to each other, you could be forgiven for thinking that they had spent years working side by side. The conversation was free-flowing and easy, though not without an edge. Gillian, Peter, and Dick have all spent large parts of their careers in institutions that are either skeptical of the synergistic approach to healing, or hostile to it. The documentary, for Gillian especially, was not simply about exploring Dick's work and bringing his ideas about healing to a wider audience: It was also an implicit critique of the rigid orthodoxies of mainstream psychotherapy. She was concerned about how exactly to strike the right balance between speaking truth to power without alienating the very people she was trying to reach.

In what is perhaps an act of journalistic negligence, I had left my notebook in my bag. By the time the evening wrapped up, I was too tired to take any further notes. Instead, I lay awake in bed, waiting for sleep to come, unused to the silence of the woods—which is to say, unused to the sound of the water and the wind, the creaking of the pines, the midnight world of the lake.

Part 4: The Beyond that's Within

Over breakfast the next morning, Dick shared an insight with Gillian, Peter, and me: "This morning, when I got up and looked out the window, I realized it wasn't me greeting the sun, the sun was greeting me." He sounded surprised as he said this. "I know it sounds weird to say, like, you mean you finally got it? But it's a powerful thing, to say you are being greeted by nature, you open your eyes, and they're saying hello to you." He looked out the window. The inlet was sparkling and dancing, and the sun was climbing above the trees.

"There's a spark of life in everything," Peter said.

Thus began the first real conversation of the day. Peter and Dick talked about how important it was to see that spark, to acknowledge that we are not simply observers of the natural world—we are also being observed by it. The question of what it means to see the spark gets them on to the topic of what this means for therapists. They agreed that the high-pressure nature of clinical work means that seeing the spark in every client who walks through the door isn't always easy.

"Keeping clinicians busy is a way of making *sure* they can't see the spark," Gillian argued. "The system is set up to keep therapists from that kind of deep engagement."

"And some therapists don't want to really see clients," Peter added. "So what happens when a clinician is, for lack of a better word, lazy?"

Dick mulled this over, and suggested that sparks also need to be tended. It is not just a matter of recognition but of cultivation. People, and perhaps therapists especially, need to develop their capacity for seeing sparks, and for encouraging the spark they see in others.

"Sparks are dangerous because they can start a fire," he added, looking over at the woodstove in the corner of the room. "That's why you need a grate to contain them."

"But can you really heal people with that conservative mindset?" Gillian asked.

"Yeah, maybe not."

As I observed the conversation unfold, I was struck by how well these three people, each with their own perspectives and experiences, engage with each other. There was a sense of dialectic in how they talked. A proposition is raised, considered from multiple angles, interrogated, and reimagined. This was possible because their personalities were so complementary. There is Peter, the man of action, who likes to see things in terms of practical application. There is Gillian, the skeptic, who, always being aware of the obstacles standing in the way of easy solutions, tends to present the counterpoint. And there is Dick, the mentor, who prefers to speak in parables and analogies, returning the conversation to the deep wisdom of the natural world. Hearing them talk is like listening to a musical trio. One by one, they pick up the theme, play with it, and hand it over to the next performer subtly changed.

When breakfast was over, we discussed what the day would look like. The plan was for Peter to give Dick a tour of the island. This would provide the film crew an opportunity to capture some more dynamic footage, while giving Peter and Dick a chance to talk more about synergy. I'd be following along with my notebook, observing the scene.

Because the crew wasn't quite ready yet, I took Dick aside to do a background interview, filling in some of the gaps in my own knowledge. We sat by the window overlooking the inlet, and I tried to get a sense of what, exactly, Dick means when he talks about spirituality.

"I always feel, when I discuss it, that it's very simple. Like, the tree is alive. I look at my dog sometimes, and see the spirit of that species, the total joy of them. I've never had a desire to articulate what spirituality is, aside from saying that it's the beyond that's within. Very kind of simplistic, almost, but that's as far as I would go in thinking about it."

Dick told me that formal religious faith was not part of his upbringing, first in New York City and later in Boston. But though he lived in a big, bustling city, this sense of connection with the natural world has always been part of his life. This is, of course, a very common view; sitting on Peter's Island on a balmy autumn day, listening to the waves on the rock, it is very easy to sink into a feeling of calm and connection, which is why so many people have built cottages on Nares Inlet. For Dick, it seems to go a little deeper. He has lived in the Kalahari, he has lived in a fishing village in Fiji, and he has spent time in bush camps in Alaska. He knows how dangerous the natural world can be. He is not talking about the detached, romantic feelings one might have on a leisurely holiday when he gives an example from his years in Fiji: "Low tide, high tide—it's a very spiritual thing, but it's also very practical. You're part of what's going on."

He told me a story about a sudden storm that blew up while he was living in the village of Naqara on Ono Island.

"Sometimes the hurricanes are quite severe. We knew there was an approaching hurricane, so we went into a ceremony and offered prayers, and the hurricane came up to the beach, took a sharp 90° turn up to the hill, and went there, sparing the village. I totally believe that our prayers were part of that—you know, 'Please spare our village this one time.' That's where spirituality is."

I was struck by the fact that he said "*this one time,*" which feels characteristic of how Dick approaches this stuff. It is not as though ceremonies are magical, having the power to always keep you safe as long as you do them right (at least, I don't get the sense this is what Dick thinks). It is more that there is some level on which collective human intentionality, channeled through deeply held cultural practices, can open people up to miraculous experiences. I have spent a lot of time around people who hold a more prescriptivist view of prayers (and magic), in which they are treated as a kind of spiritual Uber, a technology that guarantees certain outcomes. But the more Dick talked about his view of spirituality, the more it sounded like an expression of humble respect for our paradoxical position in this universe. We are, on a cosmic scale, infinitesimally small creatures living on a tiny planet on the edge of a galaxy that is merely one among billions; yet we are also, each of us, infinitely complex consciousnesses that contain unchartable depths, whose interactions with each other and with the non-human creatures around are laden with incalculable meaning. Spiritual language and the language of synergy are ways of holding both these conceptions in mind, both the macro and the micro, the subjective and the objective, the interior life, and the life of the ecosystem.

Part 5: Trees and Rocks

Outside the front door of the main cabin on Peter's Island, as you walk to the jetty, you pass a tree stump. The tree was cut down a few years ago after some debate in the family about the danger it posed to the buildings. It had been planted decades ago, and as Peter grew up, the tree grew with him. He still misses this tree. It provided shade, it grew by the water, and it was one of those features that we grow attached to when we know a place well.

As Peter led Dick, the film crew, and me on our tour of the island, it became clear that he knows every square inch of the place extremely well, and every square inch is attached to one memory or another. Behind the main cabin and the bunkhouse where I was staying, the land slopes up under old pine trees to a central ridge, and then slopes steeply down the other side to the marshy shoreline, which is protected by a long spit of stone jutting out into the calm waters on the lee side of the island. The island itself tapers off toward the bay, and where a third building sits on exposed rock at the highest point of land. From there, you can see the inlet start to widen out toward Georgian Bay. Peter was wearing shorts, and his feet were bare, though the ground was covered in pine needles, and the rocks were sharp in some places.

"This path takes us to a private special place to do your morning bathing," he said, leading us toward a sun-dappled meadow at the eastern end of the property. "My first memory of that was with my mother and my two siblings. We would all go together, and I didn't know the words skinny dip back then—I would just go for the dip and bring some soap and some towels."

A lot of Peter's memories of the island involve his mother. Though his family lived in Montreal, this was one of her favorite places in the world. Toward the end of her life, she began to spend even her winters on the island. She died in 2000, in a one-car accident on the highway not far from Parry Sound.

"I built this for my mother. She was told that she needed a place to store her wood," Peter said as we stopped by the woodshed. "My father built some furniture, and he was a printer. He was a calligrapher, and did all the diplomas for McGill; he'd do the Old English. He had some of that talent, and he was artistic, but I got very interested in woodworking, so for my tenth birthday, he gave me a jigsaw, and I started to play with a jigsaw and build things."

Peter and Dick started walking again, and the crew moved slowly around them, trying for the best shot, the best angle.

"Wood is so special," Peter said. "The green."

"Yeah, everybody having their own green."

"I love trees and, you know, playing with wood, working with what is. I didn't know at the time that it was a gift, but that's the sort of thing that I'm learning right now. The woods are a gift."

Peter mentioned that for a long time, it was hard for him to come back to the island. There were painful memories attached to the place as well.

"I would feel ambivalent. Do I really want to come here? And it's like, I can now. My mother died nearby without feeling any kind of ambivalence."

"Well, that's why when you asked me, 'Where would I like to stay?'" Dick said, "That's why I deferred to you. Because you want me to meet your mother. And what better way than to be in her space?"

We continued along the path, and the conversation continued along its path, until it landed, rather surprisingly, on the mid-twentieth century Harvard psychologist B.F. Skinner, most famous for his Behaviorist theory of the human mind. For Skinner, most of our behavior could be explained as a form of programming, in which we respond to rewards and punishments.

"You know," Dick said, "the big thing when Skinner first started his work with operant conditioning theory is that it was a form of rigid programming—the big thing was those things were programmed literally step-by-step. And people loved it, because doing therapy was confusing, but with Skinner, if you're not sure where you are, you just follow the program."

"I found it so boring," said Peter, "It's the evidence-based treatment, but it's so boring."

"Yeah, but you think about evidence-based treatment, which is so powerful, but the first question I ask is evidence based on what?" He turned around, gesturing at the trees. "This is evidence." The problem with the fetish for Skinner's kind of "evidence-based" approach, Peter thinks, has something to do with its focus on stripping away context. "It's surgically removed from the world, by design—let's take out all the extraneous variables. Let's take out all the context, and then we can somehow convince ourselves that these things caused this and caused that."

We continued to move along the path, the conversation about evidence, experience, and context unfolding through insights and anecdotes, until we were interrupted by the arrival of Dennis, a rough-hewn man who is the builder, caretaker, and irascible raconteur of Nares Inlet. Like Dick, he has the kind of weathered face it is hard to put an age to. Peter introduced him as the man who built the cabin I was sleeping in. Dennis was helping to ferry people back and forth, and he needed to check on some details related to timing. But he's an amiable man, and soon he, Peter, and Dick were talking about the seasonal migration of wildlife in these parts.

"The geese are gone now, and some of the loons as well," Dennis said. "The youngest loons leave right before the ice. There's wild rice all around these islands—there wasn't any wild rice here for a long time, but it's starting to come back. It attracts the geese, and the geese attract pike, because the pike feed on the geese when they're young."

It was interesting to hear Dennis talk about these things in such a matter-of-fact way. I am much more used to thinking about the land the way Peter and Dick seem to do, on the level of ideas. But for Dennis, with his prodigious knowledge of local flora and fauna, the interconnectedness of everything isn't a metaphor. It's a given fact, something he works with every day.

When Dennis returned to his business, we continued on our tour around the island, stopping periodically so the crew could get some good footage. They had a difficult task—they wanted to get dynamic footage of the two characters as they walked, but they also wanted to capture the high points of the conversation. It needed to be natural, but it also, to some degree, had to be framed—even staged. Too much direction would spoil the flow of the conversation, and too little would mean missing out on good material.

Eventually, we made our way around to the rockiest part of the island, the part that faces west, where the storms blow in from.

"The rocks are, I think, even more than the trees, what everybody loves," Peter said.

"It's easier to see the life in trees than the life in rocks," Dick mused, "but the rocks are alive. They're like buffalo coming up from the plains, before they were tilled, the buffalo rising up from the land."

"They say it's at least one billion years since they formed, these rocks. And it's rare to have them on the surface."

"So that's a gift, right? Allowing you to see us. And then what, a billion years from now, it might be again under water."

Peter and Dick stood there on the billion-year-old rock while Alice captured this series of moments on film, and Adam recorded the audio. In a billion years, as Dick said, these rocks could all be under water. Or they could have been folded back into the earth's crust, or buried in new layers of sediment. How long has this island been in Peter's family? The blink of an eye. How long will the recording of this morning's walk last? Decades, perhaps longer—another blink of an eye. The land is a reminder of the inconceivable scale of time we live within, and of how short our time on it is.

And yet, how much love Peter has for these particular rocks, which have meant so much to him. How much love all of us have for the particular things in our lives—the specific people we care for and who care for us, the specific homes we live in, our specific pain. Perhaps it is when we dwell on this specificity too much that we become arrogant and start to believe that we have mastered the world around us—that we can actually claim to own the rocks that were here before our species was born and that will be here long after we're gone. The rocks are, indeed, alive. We only think they're inert because we are too short-lived to see them move.

What would it mean to honor our own specificity and value our own worth as creatures endowed, like all creatures, with a spark, while also remembering that we are part of something infinitely larger and stranger that will always transcend the words we use to describe it? What things might be possible if we were, even for a short time, able to do this?

Part 6: Psychedelics

The afternoon was taken up with another film shoot, this time a kind of roundtable with some of the staff from Stepped Care Solutions. We talked about our impressions of the work so far, and about how this experience impacted how we think about synergy. It was strange to be on that side of the camera. You suddenly become very aware of your body, and how it might look to other people, and how your voice sounds, and whether what you're saying makes sense. As a writer, I'm used to having nearly unlimited space to organize, edit, and revise my thoughts before settling on the thing I want to say. There are so many more factors at play when you're being filmed, and obviously, we're only doing the one take. I accepted that I would need to let go of these anxieties—I accepted it, but I didn't like it. I was relieved when the session ended and the cameras turned somewhere else.

In the evening, the film crew returned to the mainland, and we set about making dinner. I'd only been there a day, but I felt surprisingly comfortable being in that space with those people. Part of it surely had to do with pitching in to prepare meals and wash dishes. Nothing dissolves space between people like suddenly needing to work together on shared tasks. Technically, I was at a client's summer home, doing work my client had contracted me to do. But it was impossible to feel that way when I saw Peter pottering about the property collecting firewood, or when I helped Gillian in the kitchen with food. We're all humans, we're all hungry, and in this space and time, that's that.

Things started to get interesting again when we sat down to eat. There was an important aspect of Dick's work that Gillian and Peter were hoping to include in the documentary, but they were also concerned about how it would play with the broader public. The conversation turned to the question of when, how, and even if the sensitive topic of Dick's involvement with the Harvard Psychedelic Research Group should be discussed.

A little context: The Harvard Psychedelic Research Group was made up of a small circle of academics, mostly from the personality and clinical psychology programs, who started experimenting with consciousness-altering substances in the early 1960s. Led by faculty members Timothy Leary and Dick Alpert (later known as Ram Dass), this group circumvented Harvard's strict rules and protocols to engage in radical explorations of belief and subjecthood through the medium of LSD and psilocybin. Following a series of experiments that would become a central part of the mythology of the sixties, Leary left Harvard, Alpert was fired, and both were transformed into counterculture heroes. Their ideas about the power of psychedelics for achieving enlightenment struck a chord with a generation that was growing skeptical of traditional Western forms of spirituality, and Leary's call for people to "turn on, tune in, drop out" became a motto of the hippie movement.

As Dick tells the story in *Indigenous Healing Psychology*, the Harvard Psychedelic Research Group was crucial to his own evolution away from the Western psychotherapeutic paradigm's emphasis on materialism, measurement, and data. "During the years of my Harvard graduate school training," he writes, "two powerful paths of consciousness emerge in my life: psychedelic research and meditation. They provide yeast to my developing insights about the limitations and mainstream psychology and, even more importantly, about what exists beyond that culturally delimited framework" (Katz, 2017, p. 33)

I'd asked Dick about all of this earlier in the day, and he'd told me about it with the wry detachment of a man who saw the sixties unfold in real time. He admitted to having mixed feelings about some aspects of the research group's legacy.

"The psychedelic group…that was pretty intense, but often misguided," he recalled, leaning back in his chair thoughtfully. "We thought we knew what was happening, but it was pretty disconnected from the realities of the world, and the struggles most people have not just to be spiritual, but to survive economically and politically. It was pretty isolated, even though ninety percent of the metaphors we were using were spiritual."

This isolation was, perhaps, partially due to the fact that the research group was spearheaded by academics at one of the most prestigious universities in the world. Though they were guided by Tibetan and Zen Buddhist teachings, and used ceremonial structures from Indigenous Mazatec *curanderismo* traditions, participants were not themselves embedded in the actual religious and cultural communities these ideas and practices came from. Dick insists on the distinction between psychedelic substances like sacred mushrooms, which have a long history of medicinal use, and hard drugs like cocaine, methamphetamines, and opioids. But he is painfully aware of how the "element of chaos" in the Harvard research group's work led to negative consequences in how their ideas have been received by the wider culture. *Indigenous Healing Psychology* contains a subsection explicitly called "Cautions About the Harvard Psychedelic Research Group."

Still, he is grateful 60 years later for the role psychedelics played in his own spiritual development.

"The psychedelics were not so much an eye-opener as a confirmation of these realities that some people call spiritual," he explained. "To me, it was another vision of the world we are in, but there are layers protecting us from seeing it, and the layers were pulled apart. It became clearer and clearer that what we are involved in even as we sit here is just one level of reality."

This realization led Dick in a very different direction from the one embraced by Timothy Leary. It led him to the Kalahari, where he learned about a radically different form of spiritual healing from = Oma Djo, and then to Fiji, where Ratu Civo taught him how Fijians use a root called *yaqona* in their healing practices. He came to appreciate the ways in which powerful plant remedies are embedded in the psychological healing practices of some Indigenous nations, and to understand that the context in which these plants are used is perhaps as important as the plants themselves. A person taking sacred mushrooms as part of a ceremony may be ingesting the same plant as a person munching down Psilocybe mushrooms alone in their home, but the experience is not the same, and will not necessarily have the same healing effect.

In the decades since the Harvard Psychedelic Research Group began its experiments, the use of psychedelics as a psychotherapeutic treatment has ebbed and flowed. For a long time, such treatments were heavily stigmatized—in part due to their association with the excesses of the 1960s' counterculture. But discussions about the use of Psilocybe mushrooms for medicinal purposes are becoming increasingly mainstream, and a great deal of research is being done into the use of psychotropic drugs like ketamine for treatment of depression. There is once again an openness to the possibility that psychotropic substances, both natural and synthetic, could play an important role in mental health.

An openness, but not an embrace: Some mental health professionals believe that controlled and potentially dangerous substances require considerably more research before they can be considered safe for medicinal use. Could including mention of psychedelics in the documentary lead to criticism of SCS?

Peter was of the opinion that simply talking about psychedelics in a documentary could hardly have negative repercussions for the company, but he was also aware that he was part of a team, and must take into consideration the concerns of others in leadership—even if he thought they were being overly cautious. Gillian, always sensitive to how institutional dynamics can smother good ideas, transforming even the most radical vision into empty rhetoric, agreed. Where their perspectives differed was in their lived experiences. Peter himself admitted that as a White man, he was more often given the benefit of the doubt, and grace when he failed to achieve something. But Gillian knows that when Black women try to push for systemic change, it's often one strike and you're out.

It became clear to me, as an observer, that SCS was in the middle of navigating a significant challenge. Over the past few years, it had become a successful company, which attracted the interest of powerful governments and organizations. What was once a scrappy start-up was turning into a medium-sized business with a stake in maintaining and expanding its influence. How do you keep the company pushing forward without succumbing to inertia or making so many strategic compromises that the mission is watered down and forgotten?

At first, Dick simply listened. But as the conversation became more heated, with Peter and Gillian speculating about the various people who might be uncomfortable if the documentary were to feature a discussion of psychedelics, he interjected. "Peter, if you're talking about these people so much, you should probably be talking to them." Dick remained as calm as ever, but his eyes had a focus and an attention that was electrifying to watch. I became aware that I was in the presence of a man who had spent decades working as a healer—sometimes in very stressful situations, where there was a lot of pent-up emotion.

"Yeah, you're right," Peter said, and I realized there was another dynamic in the room. Gillian and Peter are both a generation older than myself, and I tend to see them as people who, after long careers, are achieving the height of their power and influence. They have spent decades developing the skills and wisdom needed to build something vital and durable, something that can concretely change the world for the better. What more could they have to learn, at this point? But they are still at least a generation younger than Dick—who did, after all, mentor Peter in graduate school. The wisdom that Dick has to give is the wisdom of the Elder, who has been through all of this before, and can offer practical advice and spiritual insight. From this point on, the conversation became much more about the realistic things Gillian and Peter could do to bring people along with their vision.

When we turned in for the evening, they had a plan. On Thursday, the final day of the shoot, they would hold a talking circle in the morning. In the afternoon, if there was time, they would turn the cameras on and give Dick a chance to talk to the visitors from Berkeley about psychedelics, and the promise of plant-based healing for mental and spiritual wellness. It wasn't a compromise exactly, and it didn't allay all of the fears. But it accepted that the question would, in part, be decided by larger forces—the weather, and the subtle, unknowable dynamics of the little community that had gathered for the week.

Part 7: Healers

On Wednesday morning, shortly after we arrived at the Holiday Inn for Dick's talk, I was sent on a mission to find some plants. Sara, the director, felt that the front of the room looked a little bare, and was hoping we could put some greenery by the podium to liven things up. While the concierge wasn't looking, I requisitioned some plastic potted ferns. I don't think she would have minded. The staff at the Parry Sound Holiday Inn seemed to have a pretty relaxed energy. Most of the people coming in from Berkeley had arrived the night before, and were chatting with Peter and Dick in the lobby. One of their number, Adisa Anderson, had run into trouble with his flight, and wouldn't be arriving until later in the day, but the rest seemed to be in good spirits.

As Dick started his talk and the introductions began, I made notes on who each of the guests were, and what they specialized in. There is David Presti, a cheerful man, who teaches neuroscience and psychology at Berkeley and has a passion for student mental health. Kristi Panik, chief of psychiatric services of the University Student Health Services at UC Berkeley, is a sharp-eyed woman who has been involved in the work Peter is doing at Berkeley around the implementation of SC2.0. Chris McLean, a tall, rangy psychologist and student mental health worker who looks like he could make a meal out of the toughest hiking trails in northern Ontario, has likewise been working on the SC2.0 implementation as a way of improving student access to services.

I am surprised to discover that the next member of the Berkeley team is someone I know, at least by reputation. Tobirus Moselle Newby, a clinical social worker who also directs student mental health outreach, was involved in a previous project I worked on for SCS, and though we never spoke, it is nice to be able to put a face to the name. Gentleness, gratitude, and quiet strength radiate from him as, like Dick, he introduces himself by acknowledging his debt to others.

"I just want to give honor to three of my teachers in this life," he says, "my grandmother, Pastor Mary Howe, my granddad, Deacon Calvin Howe, and Teresa Iniguez Flores—because it is so important to honor those who walk with us and the ancestors who continue to bless us with their wisdom."

Matt Goodman, the final member of the group, is the most soft-spoken. He is the assistant director of the Center for Mental Health Innovation, and while his background is in research, he was drawn to therapeutic work because he wanted to "see the spark" in his work. "I really value data and numbers," he explains, "and I also really value my heart and my feelings and my spirit, and I feel like I'm at my best when I can have both of those parts kind of come together."

I am at first a little surprised that so many of these academics have backgrounds in the hard sciences—David in neuroscience, Kristi in medicine, and Matt in data science. I realize I have been carrying around a stereotype that synergy talk is very much a humanities and social sciences kind of language. But to think that is to underestimate the degree to which scientifically minded people can also see the spiritual side of their work, and to miss the obvious fact that the hard distinction between the sciences and humanities is a fairly recent one, and not one that would make sense in many cultures.

After Dick finishes his talk and the Q&A portion of the morning begins, the first question comes from Gillian.

"Is it possible?"

Dick smiles, and expresses his appreciation to Gillian for always bringing ideas into dialogue with reality. "I don't know how to answer that," he says, "because with me—if I'm really kind of careful, I'm not sure that I have done anything as opposed to things coming into my life. I have to be very honest about that... maybe it's just that the universe starts for us in a way, you know, and then how to be open."

Gillian continues, gently, to press on this question. Can Indigenous ways of knowing and colonial ways of knowing actually enter into dialogue? It's the question that's at the heart of all of this: the documentary, our time on the island, the value of synergy itself. Can it work?

Dick spends a little time talking about what it's looked like in practice for himself, as a White man engaging with Indigenous people. And he ends by stressing the importance of knowing the limits.

"One of the things I try to do in terms of working with the knowledge is to always respect the boundaries that were given to me in terms of the knowledge. It's so easy to appropriate… I've only been given permission for certain things, not for everything. Danny and others have given me a whole bunch of ceremonies. I don't practice any of them, except with family, because I'm painfully aware of the stolen heritages and the stolen knowledge."

But then he turns to Tobirus, with a twinkle in his eye.

"When I heard Gillian's question, 'Will it work?' the first image that came to mind is… you hold up during the Zoom meetings a sign that you made that has 'synergy' on it. You're discovering the synergy that's already there."

From here, the Q&A session becomes more of a conversation. Ideas are brought in, stories are told, examples are given. At some point, Matt turns to Tobirus and shares how listening to Tobirus talk about synergy and honoring his ancestors helped Matt himself connect in a new way with his own heritage as an Ashkenazi Jewish man.

"That's an example of synergy," he says. "I was open to how my experience of Tobirus in the moment impacted me. I also know that I had a lot of other moments at UC Berkeley that aren't like that… like bringing in the Ashkenazi Jewish part of me and people saying, no, don't bring that in here. So, I'm trying to think of how to hold both of the conditions that can promote synergy, and then also what's feasible within that environment that we're in.

"I appreciated what you were saying about the power of stories and the meaning of stories, and that sometimes in the environment that we're in, there's an overvaluing of data or numbers in an undervaluing of stories. But I also believe that the data is important too, you know, and that it's not just something extra that we give to people because that's what they're wanting, but that there's something that maybe how outcome studies are delivered and how we interpret it."

This strikes me as being an important point; I think this is how it strikes Dick, too, because it inspires him to talk about the role that difference plays in synergy. Tobirus, coming from a different spiritual tradition from Matt, inspires Matt to open a spiritual door of his own. It is not the same spiritual door, but perhaps it opens onto a similar enough place. This is one of the key differences between the spirituality Dick is speaking about and the spirituality I grew up with, which could only imagine unity as uniformity.

"The synergy thing is individual things that are disparate remain separate but respected," is how Dick puts it. "By joining together, that's where the whole becomes larger than the sum of the parts. But the parts have to be valued for who they are. And with our clients, the same thing—like, therapy is a ceremony, right? You have the counselor, and you have the client. You have to celebrate that ceremony. But as long as we see ourselves as helpers, we don't have that ability to, because our job is to *help* and to listen. And then the real challenge becomes if we just listen, is that enough? I don't know the answer to that."

Characteristically, Dick does not try to provide a solution, but instead tells a story. It is a story about a traditional Elder named Tony who ran a sweat lodge. Tony was generally a fairly quiet man, but one day Dick took one of his Indigenous clients to do a sweat, and after four rounds in the lodge, Dick turned to find that Tony was doing a counseling session with this person—not something that was strictly part of Tony's role as Elder at the ceremony.

"His role was to do the ceremony, but he was able to break out of that and connect just like we have to break out of our role as professional helpers. If we see ourselves as healers, that's liberating—because healers have all different ways of working, and they don't necessarily stay in one place."

The conversation circles back, one last time, to the question of antagonisms. How does one respond synergistically to the very structures—the colonial structures, the bureaucratic structures—that would seem to make synergy impossible? Dick gives an answer that feels almost like a gentle form of trolling, an answer that reminds me of how common trickster figures are in folktales from around the world.

"If we can respect the things that we don't value, that's beautiful because that's the generosity of spirit," he says. "In the Lakota way, there's a thing called the heyoka, which is the backward person. And Joe Eagle Elk told me never to judge a person, because they may be put on earth with the main purpose to trick you into judging them."

Part 8: Gifts

After the talk is finished, Dick, Peter, and I start the drive back up to Nares Landing. We will need to stop off on the way to pick up some money as a gift for the Elder who will oversee the talking circle the next day, and Peter asks about the protocol involved in this kind of exchange.

"All of the traditional healers I worked with refuse to accept payment for their services," Dick explains. "You'll get some healers who charge, especially in the city, where people are more removed from their traditional communities. And problems sometimes come up around that. You know, not everyone who says they're an Elder or a healer has a community that would recognize them as such, the 'plastic shaman,' and all that. Communities that are out on the land—sometimes the gift might be part of a moose."

He goes on to concede that in places where wealth and bounty have become synonymous with the money economy, money is acceptable, so long as it is given as a gift. This strikes me as being a curious and somewhat counterintuitive distinction, but Dick says that a "gift" isn't really about what is given. The problem with payment is that it presupposes that every given thing is "owned," and that by exchanging money, the ownership passes from the proprietor to the purchaser. Because the ability to heal is itself a gift, asking someone else to pay for healing is a kind of offense against the spiritual world from which these things come. At the same time, healers need to be thanked for sharing their time and their energy, and this thanks should come in some sort of material form.

"You should give the gift before the ceremony," Dick says, "so that it is a way of showing appreciation. Don't hang on to it until the end to see if you got your money's worth."

Implicit in this form of gift exchange is the idea that one gives from what one has to give. The logic of payment, in which a specific monetary value is attached to goods and services, creates a world in which healing becomes readily available to the wealthy (who may not appreciate it) and is withheld from those who cannot afford it (however much they may need it). To those who are used to thinking of everything in financial terms, gift economies can seem somewhat exasperating. How do you figure out what an appropriate amount to give is? What is the bare minimum, and how much is too much? But these complexities are part of the point: Rather than checking a list of prices and choosing whatever services you want, you actually have to think about the value of what you are receiving. You also have to enter into a relationship with another person and consider their needs and desires.

The argument one hears against gift economies, at least in the context of the industrialized, capitalist world, is that they are inefficient and irrational. One of the bedrock assumptions of small-l liberal economics is that the pricing system is the most scientific way of determining value, and that putting a price tag on something is a way of representing the exact amount of labor and resources that went into producing it. Doctors, for example, can charge a large amount of money to treat patients because they spent decades of their life studying medicine, and paid a lot for the privilege of doing so.

And yet, as we board the Pretty Ugly at Nares Landing and start our trip back to Peter's island, I am struck by how deeply bizarre this way of thinking is. It is obvious that the little piece of land Peter's family has been living on for a century has inestimable value to him. And if it is of inestimable value to him, how much more value must this entire stretch of coastline have for the Anishinaabe people who have raised families here for millennia? Capitalism's great trick is to fool us into thinking that the importance we place on our homes, on the trees we grew up playing under, on the wild rice that feeds the geese, is merely "sentimental." But aren't all the things that matter to us ultimately sentimental? Isn't human sentiment at the root of the very concept of "value"?

By boiling everything down to a number, the market economy I have spent my entire life living under has, absurdly, determined that cutting down trees to create more land for grazing cattle is more important than maintaining these trees as a carbon sink. And though we have known for decades that this kind of environmental destruction is driving rising temperatures that have already led to the displacement of millions of people, beef is still affordable enough for billions of people to eat it every day. And it is about as expensive to fly back and forth across the Atlantic Ocean as it is for me to live for one month in an unheated apartment in the north of England, despite the massive environmental costs associated with international air travel and the relative shabbiness of my abode. The capitalist way of calculating value simply does not seem to account for the long-term impact of our rapid technological advancements. Beneath the icy rationalism of the capitalist system of value lies a Lake Baikal of world-historical lunacy.

Part 9: Fire

Dennis's boat is much larger than the Pretty Ugly. It has comfortable banquette seating, a cabana on deck to keep the rain out, and a stereo that's playing classic rock hits from the previous century. I suppose that when you spend as much time out on the water as Dennis does, you like to travel in style. You also like to tell stories, and Dennis has a lot of stories. The best one is about a trip he took with some of his friends one winter, driving their snowmobiles hundreds of kilometers out over the ice, all the way around Georgian Bay to Manitoulin Island. This, to me, sounds about as crazy as choosing to swim the English Channel. But Dennis talks about it as if it was no big deal. He grew up on the lake. In the summer, water is a conduit connecting disparate islands and communities. In the winter, the ice turns it into a road—at least, it does for those like Dennis who know its ways and understand how to navigate them safely.

Dennis has an encyclopedic, almost stupefying knowledge of this stretch of coastline and its history. He talks about the rise and fall of the fur trade, and how money could still be made by trapping when he was younger (now, of course, furs are considered very un-PC). He talks about the cottages we passed, so many of which he has built with his own hands. He points to the large plate glass windows on one particularly large new summer home, and explains they needed to be brought in by ship, all the way from Parry Sound.

We pass by Peter's cottage, and now we're in a part of the inlet I haven't seen before. We're heading out to an archipelago where the narrow channel joins the great expanse of Georgian Bay. The plan is for the film crew, some of whom are traveling with us, to shoot interviews with Gillian and Peter, while the rest of us make a bonfire and have some food. The island we're going to is a favorite spot for barbecues, apparently: Dennis says he'll be bringing his own family out here the following weekend for Thanksgiving.

When we reach it, I can understand why this island is a good spot for a fire. Out here, on the edge of the bay, the ancient rocks are scraped clean, with only a few stands of small cedar trees at the crest and little pouches of grassy marsh in the hollows. It's like walking on another planet. I try to make myself useful by gathering up some driftwood, but the only branches and logs I can find are so bleached and aerated they feel liable to crumble in my hands. I keep looking. After spending so much time around people, it's nice to be out here, with the golden hour approaching, listening to the waves on the rocks. I grew up visiting the south shore of this same bay every summer. There is a part of me that will always feel that these barren rocks, these twisted cedars, are the most beautiful landscape in the world.

Back at the fire pit, people are chatting in little groups. Typically, in this sort of situation, I've found that people tend to stick with the people they already know. But there is a pretty free intermingling of film crew, Berkeley, and SCS folks. Chris, the crew member whose family also has a cottage on Nares Inlet, has his camera out. The light is amazing, and he's taking portraits and group shots against the background of the lake.

I end up in a conversation with Sara Alfaro, Bella, and Mike Hachey. They're talking about the film industry, and in particular, how strange the current creative climate is. Sara, whose ancestors hail from Latin America and Iran, has found herself suddenly offered a number of opportunities to direct projects that deal with diversity and racism. And she sometimes finds herself wondering what it means that *she* has been given this work. Is it because she is herself "diverse" and a woman? Is it because she's a really good director who knows how to tell a story well? For the past ten years, the culture industries have been consumed by debates about representation, and equity, and which stories are told and who gets to tell them. But however well-meaning these debates often are, the underlying structures of film and television (like literature, like music, like all of the arts) are still run by a lot of the same people and interests that have always been running them.

A little later, Bella tells me that she feels the same way. She explains that both of her parents live with forms of disability, and the support they are provided by the state isn't close to being adequate to their needs. As a writer and a director herself, she wants to do a documentary about disability rights in Canada. These stories need to be told, but telling them isn't always easy. There are certain narratives people expect, and filmmaking has a bad history of reinforcing pernicious ideas. How to create something that honors the lives and struggles of individuals, tells the whole truth, and empowers rather than disempowers? These conversations are the kinds of conversations I have been having for years, because they are conversations that any thinking person who wants to make art needs to be having. Real talk about the hard truths of making art under capitalism is the only thing that helps me move forward in my own work, and it's nice to share this with Sara and Bella because they are in the thick of it as well.

It's obvious that there are parallels between these conversations and the conversations between Dick, Gillian, and Peter I've been listening to all week. Whether it's the culture industries of mental health, maybe we're all engaged in the same struggle. Goodness only knows I have often needed to ask myself how far I'm willing to sell out my principles to publish novels or write essays. The temptation to just go where the money and power are and keep your mouth shut can be overwhelming.

In the interview she shot on the island, Gillian is composed and professional throughout. She is careful about what she says, and focused on communicating her ideas in a way that will be accessible rather than alienating. But when we talk about it over dinner, some of her frustration comes to the surface.

"The entire Western world was shaped by oppression and injustice," she explains, "and that has infected every part of our systems, including the mental health system. So I have to assume, as a protocol, that anyone working in those systems, consciously or unconsciously, is acting in a way that furthers that oppression—until proven otherwise."

Perhaps it is just the soft lamplight and the late hour, but in that moment, Gillian's face strikes me as the face of a seer from an ancient myth—the face of someone who has gained great wisdom and paid a very high price for it. But because I am not as wise as Gillian, and have only just started paying for the small store of wisdom I've built up, there is a part of me that wants to believe she is exaggerating. Surely, the mental health system is mostly made up of well-intentioned people who really are trying to do their best? But Gillian smiles, and repeats a line I've heard her say before.

"Malcolm X said, 'If you stick a knife in my back nine inches and pull it out six inches, there's no progress.'" She doesn't say any more, because she doesn't need to. But I know that the original quote continues: "If you pull it all the way out, that's not progress. The progress is healing the wound that the blow made. And they haven't even begun to pull the knife out, much less heal the wound."

For Gillian, healing the wound means more than expanding existing forms of mental healthcare to more people, or increasing the number of clinicians from this or that oppressed demographic. It means returning to a paradigm in which synergy is the water we drink and the air we breathe, a paradigm where healing is not bought and sold on the market.

There is a part of me that finds all of this a bit scary. I have a desire—perhaps typical of my race, class, and gender—to believe that we are at least heading in the right direction. But then I remember all the times when I've genuinely needed help and have come up against the hard walls of modern bureaucracy. Even as a White man, a member of the educated middle class, I have experienced the coldness and inhumanity, the psychic pain that comes when you realize that the people who are supposed to be taking care of you will happily let you or your loved ones die, so long as you do it somewhere else. Why pretend that just because this is normal, it is acceptable? Why make excuses for a system I didn't build?

I suspect that this fleeting desire I have to meet Gillian's skepticism with skepticism has less to do with rational arguments about "the way the world is," and more to do with the fear—and it is a very deep fear—that these systems are too entrenched and too powerful to be changed in any significant way, and any changes that the people in power are likely to make will be changes for the worse. This fear is, of course, itself a product of systems that have created an atomized, competitive, and socially violent world in which trust is constantly being betrayed. The more powerless we feel, the more quickly we will fall back on the same old hierarchies, the more desperately we will cling to what little help the systems choose to dispense, the more grateful we will be to receive any help at all.

This is the paradox at the heart of synergy. It promises a better, more human approach to living with each other, but only if we can learn to relate to ourselves and others in a way that feels profoundly counterintuitive, even dangerous, given the environment we find ourselves in. How to actually practice synergy in the context of a technologically mediated, post-industrial urban life? How to learn from the insights and embody the practices that flow from Indigenous psychology if you are not yourself an Indigenous person, if you are separated from the conditions in which those teachings were developed by hundreds of years of violent dislocation?

I do not have answers to these questions. I don't know if Dick, or Gillian, or Peter have rock-solid answers, either. But I am starting to come around to the idea that these are, at the very least, the right questions to be asking.

Part 10: The Circle

The next morning dawns gray, but the red sun still peeks through the trees to greet Dick and all the rest of us as we take the boat to Nares Landing. The rain looks as though it will hold off until the afternoon. Perhaps we'll be able to hold the talking circle outdoors after all. The journey to the mainland has, over the past few days, become as familiar as my daily walks to the grocery store back home, and I marvel for a moment at how quickly a place can get under your skin. The rocks and islands that loomed out of the darkness on my first trip out have definition and texture and meaning. This isn't just an anonymous stretch of coastline to me anymore.

When we reach the shore, we split up and travel in separate vehicles. Dick and Peter need to pick up Marilyn, the Elder who will offer the welcome and oversee the formal beginning of the ceremony, while Gillian and I go directly to the rented cottage where the talking circle will take place. The plan is to hold the circle in the late morning and early afternoon, and then have Dick lead the session about psychedelics and plant-based healing afterward if there is time. In the car, Gillian talks about the anxiety she's still feeling about this. The conversation we'd had the other night hangs in the background. I'm not sure what I could say that would alleviate her fears. I'm not sure it would even be right to try to alleviate them. She's been navigating these currents for a long time, and she knows far better than I do how important practicing caution can be.

The cottage has a bustling, festive atmosphere. The film crew has set up a command center in one of the rooms, where they're going over footage from one of the previous days. Some of the crew members are carrying cameras and microphones around, while others are chatting with the Berkeley faculty and the Stepped Care people. A series of industrial-sized aluminum trays sit on the kitchen island filled with gray sausage patties and undercooked potatoes, and there's plenty of coffee. I take my breakfast with Tobirus Newby, Matt Goodman, and Chris McLean, who are sitting on the back patio overlooking the lake with Adisa Anderson, a member of the counseling staff at Berkeley who has, due to travel problems, only just arrived. They're talking about the working conditions at Berkeley. I'm married to an academic, so conversations about the dismal working conditions at large universities is familiar territory. But I'm still a little shocked to hear what a university as well-known and affluent as UC Berkeley is paying its staff, given that they have to live and work in the infamously expensive Bay Area.

Peter has been working there for a couple of years, and one of the reasons he took the position was to try to implement the SC2.0 framework within Berkeley's University Health Services. Student bodies have a substantial need for mental health services, and large universities sometimes get excited about SC2.0 because, at first glance, it can seem like a way to deliver services on the cheap. It is extraordinary to think that an institution like the University of California (which has investments worth $164 billion as of 2023) would be so concerned about saving a few extra dollars off of its operating costs, but it is of a piece with the same mentality Gillian has been criticizing all week, the entrenched desire of people who command resources to maintain those resources at all costs.

It is approaching noon now, and the bad weather is coming in. The Elder, we are told, has been delayed, and we won't be able to get started until she arrives. But Tara, the woman who will be leading the talking circle, is in the kitchen, and I ask her if she can answer a few questions. She is perhaps a decade and a half older than me, with short-cropped hair, bright eyes, and a kind of mournful calm about her. Her connection to Dick began when she was a student at First Nations University, pursuing her studies in psychology. She is Métis, and we talk a little bit about Métis literature—writers like Gregory Scofield, Katherina Vermette, and the legendary Maria Campbell, who blazed a shining, difficult path back in the seventies and who is still a leading light among Indigenous artists on the prairies. Tara traveled the path of academic training in mental health, but she has recently started reclaiming ceremony and identity as ways of healing. She is modest about her own work in traditional healing, and says she's still in the process of learning how it's done.

This talking circle will be somewhat unusual because she's not offering it on her own territory. The local stewards of the land are the Shawanaga First Nation, members of the Anishinabek Nation, and an Elder from Shawanaga should be present to oversee the ceremony. For Tara, this means working with an Elder she has never met before—which is not a problem, but is also not the normal way in which things would be done. She's doing this because Dick asked her to, and her respect for Dick is obvious. I've participated in talking circles before, and I know that one of the fundamental rules is that what is said within them is sacred. I will be attending as a participant, not as a reporter. Neither I nor the film crew will be recording anything that takes place within it. But I ask Tara whether it's alright if I say a few things more generally about how a circle works, and she says that should be acceptable.

And so you will forgive me if I don't go into detail about what happened after Marilyn arrived shortly before the rain, and led everyone out to the circle of chairs that had been set up on a height overlooking the lake. I will only say that Marilyn is as old if not older than Dick, and remembers many things about the land and how it is and how it has been. We sit in a circle, and one by one each of us shares something. It can be anything. The only real limitation is that you are not supposed to comment on what anyone else has shared. We go around the circle once before the rain begins. Then we go around the circle twice more inside. Each time, it is a profoundly new experience.

I have been part of circles that have contained dozens of people, and have lasted a couple of hours. But I know that circles can go much, much longer. Dick had shared a story during his talk the previous day about a circle that included 50 people and lasted two days. "They spent two days, two full days going around introducing themselves. That was the meeting. And people listened."

The experience of listening to other people share things that are important to them for an extended period of time can feel strange at first. As Dick had noted the previous day, in a Western context, introductions are often viewed as a formality that needs to be gotten out of the way before the real work of problem-solving can begin. This is done in the name of efficiency. But a rush to find solutions can sometimes obscure the deeper contours of the problem, and so the solutions that are arrived at may not be fully thought through or have the full support of the group. When the meeting is over, the issues that were ignored during the meeting have a way of coming to the surface during implementation. And so more meetings are needed. Conflicts flare up and require attention. Whatever time may have been saved by keeping the original meeting short is squandered tenfold over the medium and long term.

The approach exemplified by the circle is more straightforward: In order for a group of people to embody synergy, every member of that group needs to understand where the differences lie, and what the root causes of those differences might be. When each individual is given the space to talk about the things that matter most deeply to them, their companions in the circle have a better sense of why each person thinks the way they do. Opinions that might seem irrational start to make more sense. And because it often does take hours to go around the whole circle three or four times, if something someone else says angers you, you have to sit with that anger and think about what has caused it, rather than immediately launching into a counterargument or denunciation.

There is also a sense in which listening to other people can help each of us figure out what we ourselves think. The process of putting things into words, of finding an anecdote or tale or song that crystallizes something important for us, is something we learn from listening to other people do it. As Tara put it in my conversation with her before the circle, "Listening to other people's stories is a way of reminding yourself of your own story."

The comparisons with traditional psychotherapy are there to be made. But perhaps the crucial difference is that Tara, in the role of convening the circle, is not there to analyze or diagnose anyone, but to be an equal participant. The hierarchy that is present in the therapeutic relationship from the very beginning—of the analyst, who has knowledge and the power to interpret, and the analysand, whose words are a kind of rough clay the analyst must work with—is flattened in the circle. None of us, in the context of sharing, has any greater knowledge than anyone else.

"I was trained as a clinical psychologist," Dick had said during his lecture the previous day. "We're trained to interpret, to tell people what's wrong with them, to help, to be helpful—which is the worst thing to do. You know, we shouldn't be trained to be helpful. We should be trained to listen."

For some reason, this reminds me of a passage I read earlier in the year in an essay by the novelist Italo Calvino. "The curse of our century," he wrote in 1980, "is that every cognitive interest is transformed into an accusation…the intellectual gaze is always in search of a crime to try, a disgrace to report, a secret to violate" (Calvino, 2023, p. 291). This was for him especially true of psychoanalysis: "All it can do is investigate, look for hidden guilty motivations" (Calvino, 2023, p. 291). So often, the psychotherapist has been trained to treat their client as if the client is hiding something, and the key to successful treatment is to uncover what is being hidden. But synergy exists in the moment when the charade of mastery is dropped, and the relationship between practitioner and client becomes a relationship between two individuals, both of whom are trying to make some kind of human contact, neither of whom is equipped with anything more than goodwill and the desire to heal and be healed.

Like I said earlier, I cannot write about everything that happened during the circle. I cannot present any case studies of the breakthroughs people had with themselves and with each other. It wouldn't feel right to talk about the things I myself shared, either. What I can say is that it was exhausting, fascinating, occasionally dull, and absolutely, profoundly liberating.

Part 11: Pulling Out the Knife

In the end, the question of whether or not to have a conversation about psychedelics is addressed not through decision-making, but through the movements of synergy. We simply had too much to talk about in the circle to have time for an additional session. After the talking circle ends, the eating begins. As might be expected of a group of people who have just spent several hours listening to each other talk about deep and intimate things, there was an almost intoxicating sense of closeness. Conversations that, under other circumstances, may have taken months or even years to happen now seemed natural and welcome. I ducked outside onto the porch, where some people from the film crew were hanging out. The rain was still coming down in that gentle, autumnal way it does when the trees are still in leaf, but you can smell the change of the season coming up from the dirt. I sit down at a table with Bella. She tells me she is thinking about how strangely connected she feels to all these people she's only just met.

"Usually, I only have these kinds of conversations with my friends when they're blackout drunk. And then they don't remember anything they said the next day." The question of friendship is on Bella's mind. She tells me that she's in the middle of consciously moving away from a group of people who she is starting to realize do not share her deepest values. We talk for a while about how your relationships with other people change as you get older, becoming both simpler and more complicated. But her comment about the role drinking often plays in friendly intimacy lodges in the back of my mind. It is one of the profoundest indictments of certain strains of Western culture I have ever heard. Chemical substances have long played a role in human bonding, and the ritual of gathering around the table for a pint or a glass of wine is one thing. But what I hear Bella describe is something else entirely, something I have known intimately in my own life: The vortex of shame and solitude that spins somewhere deep inside the culture of Europe and its settler colonies, the fear of others that is also a fear of oneself, which can only find expression through a seeking for connection that is obliviated as soon as the sun rises and the fumes burn away.

Back inside, I find myself sitting next to Matt Goodman. I ask him to elaborate on some of the points he made earlier in the week about the limitations of data as a way of knowing, but the conversation soon goes in a different direction. As I had noticed when I first met him earlier, Matt is a man with a certain gravity. He speaks, and you have the immediate sense that you are listening to someone who has thought very deeply about what he is saying. He weighs his words because his words are not empty. He is also someone within whom the past is alive—the difficult but also joyful, philosophical, and capacious past of his Ashkenazi ancestors. I didn't mean to talk to him about the religious side of spirituality, but this is what we end up talking about.

Matt tells me that many of his family members would be surprised to hear that he had been able to talk so openly about how important his Jewish heritage is to him in a room full of academics. From their perspective, a room full of broadly progressive gentile intellectuals would hardly be a safe space to speak about Jewish spirituality or the ancient tradition that feeds it. He expresses the wish that they could have seen how welcoming the others had been.

There is something slightly wistful in his voice, a wistfulness, perhaps, born of the strange space in which many Jewish people find themselves with regard to the culture wars. Classed now as White, and associated (voluntarily or otherwise) with the political project of Zionism, it is not unusual in progressive academic circles to hear Jews everywhere categorized as being among the oppressors rather than the oppressed, though the Shoah is still within living memory. Coupled with the recent increase of anti-Semitism in countries like the United States, this leads some Jewish people to conclude that the political left's commitment to the liberation of oppressed people everywhere comes with a caveat: "Except for the Jews." At the same time, Palestinians and their allies remain un- or underrepresented in institutions of political power, and peaceful protests in the form of calls for boycotts, divestment, and sanctions of Israel are often met with harsh discipline in academia and government institutions. Both sides feel that the other one has all the real power. Whatever the facts on the ground, these feelings are real.

I hesitate to even mention all of this in the context of an essay about synergy. I have no special insight into the conflict between Israel and Palestine. But as a writer, as a citizen, as someone who has demonstrated for Palestinian rights, and as someone who has close friends who are deeply affected by these issues, I have publicly identified myself with one side of it—though the language of "sides" always feels slightly uncomfortable to me. I think what matters here, the reason this is important to talk about in the context of synergy, is that Matt and I sat on the couch together and talked honestly about his experience, and the experiences of members of his family, and together, we expressed our sorrow about the long history of violence against Jewish people. If synergy offers any hope at all, it is the hope that by creating a meeting space outside our bunkers, we can let go, to some degree, of our fear, and at least understand where the other person is really coming from.

This is also, I think, why synergy is so hard to achieve. Not only is fear often a rational response to a world that can be extremely cruel, it is also a part of the formation of our identities. On some fundamental level, we *are* our fears. To let go of them, even for a set period of time, is to let go of part of ourselves. Synergy doesn't mean being nice to other people, it means confronting the seductive and satisfying role that fear plays in our own psyches. To open a newspaper and see that the people we are afraid of have behaved as badly as we expected them to behave is pleasurable. It means that we were right all along. The corollary of this is that when we find ourselves (or those we consider to be part of our group) behaving badly, the desire to prove that we are, in fact, in the right can be almost overwhelming. This is a very old trap. It teaches us to see other people, who are made up of infinite paradoxes and contradictions, as having a single identity, and it keeps us from seeing the paradoxes and contradictions within ourselves. The world gets flattened out, clarified, and reduced to the most basic narratives and the simplest, oldest equation of all: us or them.

Only once we've stepped out of our bunkers can the work of charting a way forward begin. This is the work that matters, because the differences are often real. Sometimes, the differences are more than a matter of what songs you sing and what god you pray to, but involve long histories of violence and oppression that have determined for centuries who works the land and who owns it, who serves and who eats. Finding a way to pull the knife all the way out of the wound is hard. It's harder still to then salve the wound, and bandage it, and give it time to scab over. But if we say that we want healing and not harm, this is what it needs to look like.

Part 12: Synergy

Two days later, I was back in England. I flew into Heathrow Airport late in the morning, and made my way to St Pancras Station, where I waited for the train to Yorkshire. I checked the news on my phone. It was October 7. The papers were full of stories about Hamas's attack on Israel. I immediately thought back to my conversation with Matt. I don't know where he was when the news found him, and I won't presume to know what thoughts or feelings he experienced on that day, or on any of the days since. But I was reminded in the starkest possible terms that the space that had been created among that group of people was fragile. Synergy is fragile.

It takes years of care and sleepless nights to raise a child to maturity, years spent worrying about what food to make and when to be stern and when to be indulgent, and yet a single event can shatter their mental and physical health, leaving them with scars they will carry for the rest of their life. Synergy, like the trust it relies on, takes time and patience to build, and it can be destroyed in a moment. The bunkers are always just behind us, waiting, holding out the false promise of security. The real mystery of synergy lies not in its creation but in its maintenance—the small decision we must make over and over through every day of our lives to keep seeing the spark in others, to keep tending to it, to not be afraid that someone else's fire will burn our house down.

I was tired. I'd spent the previous day traveling, and I hadn't slept much on the plane. Terrible things were happening, and they were going to keep happening.

Perhaps they would happen to someone I knew, someone I loved. I opened my notebook, and looked at the passage I'd written the night before, at Newark International Airport:

> *The darkness was cool over the water as we left the jetty. In the boat's headlamp, the raindrops were shafts of white light coming in from the mouth of Georgian Bay. I crouched down in the bow, gripping a mooring ring and leaning into the storm. The waves exploded off the bow in gouts as we troubled the waters of Nares Inlet. Behind me, Gillian stood in the gap between the seats, holding onto the metal bars of the helm, a shawl tied around her face to keep the damp out of her hair. Peter stood next to her, hands resting on the wheel, leaning into the force of the boat's movement, steering us between the rocks. They stood together in the driving rain, guiding us toward the safety of the dock that we couldn't see, but that we all trusted lay somewhere just ahead of us.*

Reflection

Peter Cornish

I am, obviously, a central character in this recounting of our three days in Ontario. To offer a reflection on this chapter seems both redundant and a little self-indulgent. It reminds me that synergy is very much about context and that every figure in a story is only a small part of something much larger and more important. André sees me as a man of action, one who strives to transform philosophical and spiritual forces into something practical. While this is one of my contributions, seeing it on the page reminds me how important it is to step back at times and leave room for other things. The talking circle, as André referred to it, is a model for this. At one point that week, I had called it a "listening circle." Dick smiled, and said gently, "You know that it is a talking circle, Peter, not a listening circle." I was puzzled at first by this, until I realized that he was emphasizing the power of voices. The point was less that I had to listen, and more that others had to talk. Responding to others was not permitted, and as such the "active listening" I had been trained to do as a psychotherapist was neither welcomed nor needed. It wasn't about me, my skills in listening, or anything I had to *do*. Yes, I would be listening, but not in my action stance. I see the "I" popping up in these sentences and am reminded again that it is not about me. It is about everyone. It was a gentle reminder to leave the ego behind, put my action-oriented mission aside, and allow synergy in the circle and world around to enter the collective space and greet us just like the sun had done for Dick earlier that morning.

As this volume is a follow-up to our book on conundrums (and as a psychologist, I was trained to sense and walk toward conflict and tension), my reflections in this book are drawing from more deliberate searches for contradictions. In particular, I am tuning into opposing thoughts and feelings that would illuminate something deeper and more meaningful in my quest for well-being everywhere. One central tension I recall from the three days last year was between a sense of obligation to get something done (the creation of a video that would be interesting to viewers) and the desire to step back and trust that the universe would greet us with gifts. But this could only happen if we opened spaces to see, hear, and feel the voices and energy around us—the teachings from each other, the rocks, the water, the creatures, the land, and the sky above.

This tension grew stronger in the months following the event as we entered the post-production phase. While participants had expressed gratitude, along with a sentiment that something valuable had occurred, it was difficult to discern what that something was. And although the post-production crew created glorious cinematography that evoked serenity, it left viewers wondering what it was about, what really happened. Most disappointingly, it left some feeling kind of bored. What was the plot or the story arc, where was the suspense that would grip the viewer's attention? It didn't seem to be there. So, as a person of action, I had a problem to solve. Luckily, the stunning cinematography brought me back to those rocks, trees, and water, reassuring me that time was on our side. Our curiosity, openness, trust, and passion will get us there, and while action is a vital component, it is only one part of the larger whole. We were treading on the land lightly, drawing from it uplifting healing energy. We engaged from a position of respect, harvesting the seeds of synergy slowly and sustainably. This imperfect video was just one seed.

With the finished product still a month or so away, I am still unsure what it will bring to viewers. Despite this uncertainty, more seeds are sprouting. I see a sparkle in the participants' eyes during ongoing conversations. More ideas are blooming. A bigger, and bolder story arc is emerging. This video will be just a start, a small step in a fascinating forever journey. In looking back, we identify other seeds, such as Rachel Carson's publication of *Silent Spring* in 1962. Then we have "aha!" moments, seeing connections between historical watershed moments and the myriad of crises plaguing our late capitalist world. These connections and a sense that there is something much larger for us to work toward are compelling.

Sparks flew over a recent dinner at a quiet, modest Tibetan restaurant one night in Berkeley. We were forging connections between some of society's most wicked problems, and postulating about the turning points over the last century that led us here. What if psychedelic use at Harvard in the late fifties and early sixties had been spiritually grounded in the wisdom of Indigenous healing practices? What if deinstitutionalization and the release of psychiatric patients from large hospitals had been to holistic community-based programs that included peer support, family, and neighborhood involvement in co-design and delivery of support? What if the classic generalist-practitioner role had not evolved into the current primary care referral coordination role relying unsustainably on specialist assessment and interventions? What if psychiatry had never shifted from its philosophical foundations to a focus on pharmacological treatment? What if psychology hadn't abandoned humanistic or existential theory in favor of behaviorism and a reliance on only that which could be proven through clinical trials? What if pain management had been holistic? Would we have an opioid crisis? What if mental health treatment included a focus on environmental stewardship and land-based activities inviting neighborly and ecosystem connection simultaneously? Synergy was boiling around us that evening. This reassured me that my more tempered action orientation was bearing fruit, allowing something larger to evolve and flourish.

References

Calvino, I. (2023). *The written world and the unwritten world*. Penguin Modern Classics.
Katz, R. (2017). *Indigenous healing psychology: Honoring the wisdom of the first peoples*. Healing Arts Press.

Chapter 6
The Education of a Healer: Recovering Indigenous Practices of Synergistic Healing

Gillian Berry

What Is a Healer?

The desire to seek out alternative approaches to healing led me to immerse myself in the rich traditions of African philosophy, and what day-to-day relevance these traditions might have for the healing professions. I enrolled in a PhD as I felt my training had not equipped me to navigate the specific challenges that faced people of color; I wanted to know how Black people have maintained their mental well-being during periods of great difficulty in the past. How do Black people maintain their mental well-being in the present, even in places where the social structures are hostile to Black cultures and self-expression? But as my studies progressed, my PhD supervisor at the time suggested I wouldn't be able to find the answers if I stayed in England. I could not write about anything to do with Africa without spending time on the continent—and not just spending time there as an academic researcher, but actually going there to live.

I had originally planned on staying in Zambia, where I had friends, but circumstances led me to Botswana, where I took up a position at the University of Botswana in 1999. After a long and complicated journey, I flew into the airport at Gaborone, and was greeted by a friend of a friend, someone I had never met before. I was welcomed with open arms. I couldn't believe it—these people didn't know me, and yet I was embraced and immediately shown what to do. I was given a key to someone's house and told to make myself comfortable. They were at work, but they'd be back later on in the day, and in the meantime, I was to eat whatever I wanted and have a bath if I wished. I definitely wasn't in England anymore.

As I settled in, I got started on the research I'd come to Botswana to do. My goal was to study traditional healers to see whether African approaches to well-being

G. Berry (✉)
Stepped Care Solutions, Mount Pearl, Canada
e-mail: gillian.berry@steppedcaresolutions.com

P. Cornish, G. Berry (eds.), *Synergy*,
https://doi.org/10.1007/978-3-031-88493-1_6

might hold a key to how Africans in the diaspora can survive and thrive in spite of the obstacles they face.

Traditional healers in Botswana are part of a rich Indigenous tradition that has survived for centuries. Even as colonialism, war, and urbanization have brought chaos and disruption, this tradition has survived, adapted, and continued to serve people (as have so many Indigenous healing traditions around the world). Though these practices are often maligned by some Westerners, who see them as "unscientific" or "primitive" and the practitioners as "witch doctors," the reality of traditional healing in a place like Botswana can't be understood using such simplistic and racist caricatures. It needs to be seen as a living tradition with deep roots in ancient African classical civilization that has been passed down through oral history, written and living practice texts, paradigms, worldviews, and values.

As an African-centered researcher, acculturated in a Western scientific approach, I felt it was normal to research a question I had. But my question was more than just seeking an answer for an answer's sake—it was personal. I knew that my task in Botswana was neither to impose my own views on the people there, nor to pretend I could be a neutral observer. In order to understand the perspectives, practices, and philosophies of the healers I would be working with, I needed to participate in a dialogical experience with them. I believed that researchers who were only motivated by curiosity, or racial pride, or the desire to add to a body of scientific knowledge would not be able to discover the embodied, experiential truths that are valued so highly by African civilization, in which consciousness, or awareness, is regarded as being the ultimate goal—with self-knowledge being considered as the basis of all knowledge. An African-centered perspective aims not just to understand what is being observed, but to offer corrective measures—to "pursue the path of liberation," as Bekerie (1994) put it. In studying traditional healers, I wasn't just trying to learn more about their practices. I also wanted to understand how their wisdom could apply to my own life and work. Perhaps I could even find the "mystical" concept that was the "Black perspective." I was not looking for data; I was looking for transformation.

This meant that in pursuing my research, I would need to interpret gestures, comments, feelings, encounters, and observations as information that could aid my understanding of an African-centered perspective on counseling communicated via the spiritual world. I would need to spend time in the world of traditional healers, on their own terms, let their paradigm become my own, and conceptualize the extent to which these practices could be incorporated into clinical work. To fully immerse myself in the mindset and approach of traditional healers, I contacted the Association of Traditional Healers to ask if they knew anyone who might be interested in talking to me.

And that's how I met Kitso Mbenge.

A Ringing in the Ears

Kitso was a healer. The first time I called him on the phone to explore the possibility of meeting, a pleasant voice answered the call, and engaged in conversation as if he had known me for about 20 years. He had no hesitation about being “a participant in my study,” and we arranged to meet. He accommodated my teaching schedule by agreeing to meet on a Saturday. Rather than seeing this as an inconvenience or as a favor to me, he was excited to be able to introduce me to his wife and children.

“I didn’t expect to see a Black person,” he said, as soon as the usual introductions were over. We then spoke at length about my British accent and the Dutch influences on his language and behavior (he had once lived in the Netherlands). He started out by giving me a tour of his home and introducing me to his wife and young children. I immediately felt embraced by his hospitality. It was a way for him to tell me who he was: If I wanted to learn more about the work he did, I needed to have a sense of who he was as a person. There was a seamlessness to it, which felt so unlike what I was used to back in England, where divisions between public professional life and personal private life are so rigidly enforced. From the very first moment, I was seen, respected, appreciated, and valued not just as a visiting researcher who wanted to ask questions but as myself. The trust I felt as a result was palpable. My defenses came down, and perhaps unconsciously, I mirrored the trust he was showing me by opening up about my desire to learn for my own mental well-being. I would realize, upon later reflection, that I had never been able to open up to the numerous clinical supervisors I had had, regardless of their skin color, because I hadn’t felt it was safe to do so. Kitso’s way of introducing himself made me wonder if I could build trust with my own clients more quickly if I were simply more open about who I was—or would such openness be misinterpreted as being an inappropriate closeness with a client, a matter of ego, or a breach of professional ethics? All of this reminded me that I needed to take a very humble approach to my work. I needed to respect what he was showing me, and to enter his world and his way of working on its own terms, rather than judging his approach against the Western metrics of objectivity and rationality I had encountered at university.

When the tour of his home was finished, and we had eaten, I turned my tape recorder on. There were many questions I wanted to ask him, but we began by talking about how he had become a healer. This is the story he told me.

When Kitso was a young man, he wanted to be a graphic designer, and he left Botswana in the early 1980s to pursue his dream in the Netherlands. But a strange thing happened when he landed. He experienced a ringing in his ears that wouldn’t leave him, and on his very first night in the Netherlands, he had a dream in which he received a very clear message that he needed to go back to Botswana and become a traditional healer. “I had no idea what was going on,” he told me. “I was brought up in a Christian family and was not involved at all in ‘those types of things.’ But the dream had been so clear.” The next night, an even more explicit dream came to him: He saw the bones and the whisk he was supposed to use as a healer. He stayed in the Netherlands for almost four years, but the dreams never stopped. If anything, they became more explicit. "I had no idea what it all meant, but I knew somewhere deep inside of me that I should return to Botswana," he recalled. Kitso resisted the dreams for almost four years, but when he finally went home, the ringing in his ears stopped, and he asked his parents to take him to the Elders in the mountains.

"They both thought I was possessed and crazy to be talking about all these things," he said. "They didn't believe in the traditional healing process, so it was difficult to convince them. They ignored me and abandoned me, and so I went alone to the mountains. I was returned and told I was too young to go alone. I had to go back and have some elderly person to accompany me to the place." Eventually, Kitso's grandfather agreed to bring him to the Elders. This grandfather had sought out traditional healers when living in South Africa, and he had come to believe in their power. He was the only one in Kitso's family who supported him on his journey.

The process of becoming a traditional healer in Botswana is very different from the process of becoming a doctor or psychotherapist in the West. When Kitso and his grandfather arrived at the home of the healer he would be studying with, he was one of six initiates who had been instructed to come. They had been told to bring gifts and animals for a purification ritual, as well as songs. The initiates would stay with the Elder for as long as it took for them to be deemed ready to practice. Training and readiness are both individual matters: There is no set curriculum or standardized exams to determine who is worthy of becoming a healer. One's training depends on one's gifts and character, and one's readiness is judged by the Elders, the community, and the spirits. "I thought it would be at most a couple of months," Kitso recalled, "but those months turned into years. How long you stay depends on your spirit. It is they who dictated. People danced until they reached a trance, and that is when people say the spirit has come."

The training itself was extremely rigorous. The initiates lived together in a house and woke at 5 a.m. every morning to begin the day by singing and washing themselves in the cold water of a nearby stream. They were responsible for their own food and needed to purchase it with money sent by their families or forage wild vegetables. Living with the other initiates was a crucial part of the training, because they could support each other in their studies while also sharing household chores and other common goals of daily life. By living, learning, and working together, the initiates were being trained to engage collaboratively with each other. This sense of lived community is no longer part of Western education, which is increasingly conducted in an atomized, individualized way. Perhaps if we designed the training of our healers to require this kind of collaboration around daily work and study, they would be better able to engage with the diversity of the general public.

Dreams were a big part of their training, and they took traditional medicines to stimulate visions. Every morning, they would tell the Elder what they had seen over the course of the night, and she would provide guidance. While life at the trainee house was austere, Kitso remembered the dreams as being one of the most difficult parts. There was a problem with his dreams, and eventually, the Elder determined that she was not the right person to guide him. He needed to learn from another Elder on the eastern side of Botswana, by the red mountains, where there was a larger training center. The laborious process had to be started again under new guidance.

To some readers, the importance of dreams and the spiritual dimension in Kitso's training might seem difficult to understand. It is a clear departure from Freud's interpretation of dreams as a means for an unconscious wish or symbol to become known, and it clashes with the Western materialist conception of the world, where decisions should be made based on quantifiable evidence, without reference to spiritual concepts or practices—which are viewed as being at best idiosyncratic, and at worst completely untrue. But these are the very things that lie at the heart of traditional healing. Kitso's world is one in which the spirits are real, and the ancestors are always watching you. In order for the healing to work, one must accept these facts and take seriously the worldview out of which they come. The healer and the patient must be open to the possibility that healing requires the active participation of the spirits, which respond individually to each individual person, meaning that the results will be different for each person. This cannot be measured by a researcher using Western methodologies, but must be experienced in a spirit of openness and trust. The Indigenous healers from whom Kitso learned his practice were aware that they would be held to a very high standard by the community, and by the spirits themselves. Their ethical guidance did not come from an external board or authority, but arose from their internal sense of value. This is one of the reasons why dreams played such an important role in the education of healers. Learning to interpret and listen to dreams meant being open to the voices and judgment of the spirits.

From this perspective, a mental health practitioner adhering to the code of ethics or being intentional about applying its principles in one's day-to-day practice isn't enough. A traditional healer answers to spiritual powers that are beyond them, and understands that they are accountable for everything they do. Trust, again, seems to play a central role in their work as it relates to both the seen and the unseen worlds.

At the second training center, Kitso was told that his problems were the result of his long sojourn in Europe. He should have started his training years earlier. But he was welcomed into the community and settled into its rhythms. The chief healer had five wives, and three of them were healers and trainers responsible for the initiates, who were a diverse group—men and women, old and young. As in the first training center, dreams were at the heart of the process, and initiates needed to seek out a person who was having the same dreams as them. In Kitso's case, this was the daughter of the chief healer, who was also a trainee.

Life at the new training center was no less rigorous than it had been at the previous one. Every day, initiates dug for medicinal plants and provided healing and protection to the people of the nearby village. For 24 days a month, they would dance every evening. Sex was prohibited, and initiates were expected to be completely honest in all things. Pretense was punished severely, and initiates who were faking some aspect of their dreams or spiritual experience were quickly exposed and sent away. The training process was not simply about teaching the initiates what medicines to use or how to understand their dreams; it was also about weeding out initiates who were there for the wrong reasons and were not going to serve the people well. It was understood that healing medicines are incredibly powerful, and healers themselves will become powerful members of their communities. Readiness was not simply a matter of knowledge, but of character formation, and this could not be separated from spiritual maturity.

When a healer has completed their training, there is a test. The initiate will return home, and before they arrive, their family will select a cow. If they can identify the right cow, it will be slaughtered. Its skin will be turned into a mat, and its horns and bile will be used as the spirits demand. Gifts will be offered in celebration, not to the initiate, but to the spirits. If the initiate is not able to identify the cow, they will be sent back for more training. Because so much of traditional healing is about discernment and being able to listen to the spirits, this is essentially a test of whether or not you have the awareness to take on such an important role. It is not only a test of your knowledge, but of your ability and wisdom. This is important, because the forces traditional healers are dealing with require real character if they are to be used beneficially.

"Some people use it for the wrong reason," Kitso said. "If your heart is bad or you are mixed up. The spirit goes with your behavior, and you have to be truthful. If you are not, then you get involved with things that are not good for people."

In pre-colonial Botswana, a clear distinction had been made between forms of spiritual healing that were good, and the "witchcraft" that is used to harm. But under British rule, all of these spiritual practices had been bundled together under the Witchcraft Act of 1928. The colonial ways of thinking imposed during the occupation were deeply at odds with traditional wisdom, and they have had a lasting effect on Botswana. For example, the colonial conception of time, based on clocks and calendar months, meant that people were forced to think about time in an inorganic, regimented way. In Western-style mental healthcare, for example, clinicians often prescribe a course of treatment for several weeks, regardless of how long it takes the client to show signs of improvement, but in traditional thinking, time was more fluid: Treatment, like training, lasted as long as it needed to last, and was judged not in terms of months but in terms of internal development. Colonial conceptions of value turned everything into a commodity, with money as the ultimate arbiter of how much a thing was worth. In all cases, colonialism imposed strict binaries, with Western ways of thinking universally conceived of as being "good," while Indigenous ways were conceived of as being "primitive" or "evil." Every aspect of the well-being system had been impacted by colonialism, and Kitso believed this had been profoundly damaging to how people in Botswana thought about traditional healing.

"Some people have gone on television and said publicly that traditional healing and witchcraft are the same. Even academics have said the same thing. People are saying so many wrong things about traditional healing, and it's a shame. Traditional healing really works, and it has helped African people for a long, long time."

When explaining how a traditional healer approaches someone with an ailment, Kitso gave me the example of a plant with many different uses. Its root might be made into a drink that can cure high blood pressure, or mixed with milk to become a laxative, and the very same plant might be poisonous if not cooked properly. But there is another dimension to the healing as well: "How you are, your spirit, adds to the roots." For Kitso, the spiritual world has a role to play in the healing as well, and if the correct rituals are not followed, the medicine can have negative effects. Good spirits can help facilitate healing by working through the healer, while bad spirits can cause harm. Unlike mainstream medicine in the West, with its distinctions between body, mind, and spirit, traditional healers in Botswana understand the world holistically. For medicine to work, it must be properly aligned with the spiritual dimension in both the practitioner and the client. A healer who is not operating with good intentions can do damage regardless of whether or not the medicine is the correct one for the ailment.

This is one of the reasons why Kitso's training was so rigorous. Not just anyone can do this work, and the Elders understand that their role is to prepare initiates who are ready to face the spiritual and intellectual demands the work will place on them. How different this is from Western professional education, which operates as a for-profit business, and judges students solely by their intellectual capacities based on a series of tests and writing assignments, and imagines that healing can be done in a one-size-fits-all way!

Kitso acknowledged that there are complications involved in adapting traditional healing to modern conditions, though, and expressed concern about how the economic incentives of the market were changing the way healers worked. "Traditional healing was never a commercial enterprise," he explained, "but the conditions of today mean that they must charge, as they have to go and get medicine, they have to use transport, they have to pay rent. But the people are used to cures for free, or to give a gift of a cow…But today, some healers charge high prices because they want to make money. So they are spoiling it for the true healers." In his opinion, this was another problem that had its roots in the colonial past. The experience of colonialism had corrupted traditional practices on a deep level, introducing the concept of money into the system. Before colonization, the emphasis was not on growing wealthy but on spreading healing; the financialization of the economy introduced a profit motive that had not been present before. And because there was no legislation governing traditional healers outside the Witchcraft Act, it was hard to ensure that people who wanted to be traditional healers were getting the right training and practicing ethically. But despite these headwinds, the tradition lived on. Kitso's own practice was proof of that.

As he told me his story, I was struck by how much it contrasted with the assumptions that underlie Western training and education. From the first moment I met him, Kitso had been teaching me to see healing differently. He had not decided to become a healer; he had been called. He had not simply studied techniques; he had learned to embody them. He had not just paid a fee to receive a service from his teachers; he had been mentored by them 24 hours a day. He had not passed an exam; he had needed to prove that he had the spiritual maturity and insight to do the work. Being a healer was not a role he played between the hours of 9 a.m. and 5 p.m. on a weekday, a uniform he put on in the morning and took off at the end of the day; it was *who he was as a person.*

Kitso's training and practice as a healer, in short, were synergistic. He understood that he was part of a whole—the universe—that was much larger than himself, and his existence had a purpose. He had a purpose in life that went beyond what he personally desired, as he deduced when he went to the Netherlands and experienced the ringing in his ears that was a sign he was avoiding his calling. Failing to live out this purpose, or failing to measure up to the ethical standards of the community, would be judged in the next life. Like everyone else in the community, Kitso understood that the ancestors were watching him, and there would be consequences for his actions in this as well as the next life. He had learned about the spiritual world, and had learned to understand dreams and the rich symbolic reality they drew on, and he had learned how to use medicinal plants, but this was not, arguably, what made him a healer. It was the fact that he listened to the ringing in his ears and obeyed his calling, a calling that had been affirmed by his Elders during his strenuous years of training. This is what allowed the healing energy to synergistically flow through him and into others.

Reflecting on my own training and education, I am struck by the fact that I was never asked to tap into the healing energy that flowed from my own unique identity and calling. In fact, the unwritten rules seemed to suggest the opposite—I needed to repress and ignore what was special about myself and conform to the rules that were laid down for me by the system.

Applying the Lessons of Traditional Healing in a Modern Context

I finished my research at the University of Botswana in 2003, and decided to move to the United States, where I got a job working in Maryland's public child welfare system. My experience in Botswana, and my conversations with Kitso Mbenge, had given me a profound sense of what synergy can look like. I wanted my helping/healing practices to be just as impactful, but was applying a synergistic perspective possible, appropriate, or legal in a highly bureaucratized social work context in the United States?

During my first three months on the job, I was given the case of a foster child who needed to be taken to the doctor. I had picked her up from her foster family and was driving her to the doctor's office when she pointed at a house we passed and told me that her family lived there—meaning her birth family. It is well-established that many foster children are really helped by having some connection to their birth families (often, the first thing they will do when leaving foster care is seek out their birth parents or siblings), so I pulled the car over, and we paid them a visit. We went in, sat down, and the girl had a chance to talk to her siblings for a while before we continued on to the doctor. By the time I dropped the girl off with her foster family, her birth mother had lodged a complaint about my behavior. How dare I bring the child to the house unannounced!

I understood that what I had done was against protocol. But I had done it because I saw it as being in the best interests of the child, who, after all, was the person I was there to help. My approach was to be flexible, and to try to figure out what would be in her best long-term interests. But of course, the birth mother was right: I had broken protocol, and I needed to apologize because I had acted in a way that the system did not want me to act. I was struck, however, by the fact that nobody was interested in having a conversation about whether what I had done had been *helpful* for the child. Her interests didn't seem to factor in. Had I known that the route I was taking to the doctor's office would take us past her birth family's home, perhaps I would have taken a different route. But as I saw it, whatever damage might have been done by giving her a chance to spend time with her siblings far outweighed the emotional turmoil that would have come from telling her we couldn't stop, that she didn't live there anymore, and we had to keep to the schedule. If there was a chance her family would be present in her life after she aged out of the system, should a relationship with her siblings and birth family not be nurtured?

For me, this case served as a reminder of how difficult it is to practice synergistically within the context of a system that is not set up to follow the principles of flexibility, agency, awareness, creativity, and empathy that are so important for synergy. As so often happens within bureaucratic systems, rules—even rules that are meant to protect people—become an end in themselves. Rather than judging whether or not this or that action is the right one to take for a client in our care, we argue about whether it conforms to the letter of the law. In our attempts to build a system that is general enough to meet the needs of everybody, it ends up meeting nobody's needs but those of the system itself.

What would a more synergistic solution to the above case have looked like? What if, instead of appealing to the protocols, an attempt had been made to engage in truth and reconciliation between all of the people involved? What if we could have put the girl's needs at the center of the situation, rather than treating her as an afterthought?

I don't know how the situation developed in the wake of this event because, not long after, I was promoted to a supervisory role (primarily, I think, because I was the only licensed candidate). This also proved to be a learning experience, as I was now responsible for helping to direct staff in ways that would encourage them to be more synergistic in their own practice. Here, too, I saw how strict adherence to the rules could get in the way of delivering the kind of care we were ostensibly there to provide.

One of the social workers I supervised, who was particularly legalistic about the rules, had a complaint filed against him by one of his clients, a single African-American mother with seven children. She was clearly struggling, and had been around social services long enough to know how to manage the system. She called to let me know that she didn't like the social worker. Complaints such as this need to be investigated, and I asked the social worker what had happened. He told me that she hadn't wanted to sign the service plan, which, for legal reasons, must be agreed to by both parties. I needed to get the client's side of the story as well, and I could easily have just asked her to come by the office. But instead, I opted to visit her in her home. One of her children answered the door, and I kneeled down to make eye contact with the child, and engaged in a conversation about their day. The mother immediately told me that the problem with the previous social worker had been his refusal to engage on a human level with her or her children. This, it turns out, was why she hadn't wanted to sign the service plan. She had felt disrespected and belittled, as if she and her family were just another problem he needed to manage.

On the spur of the moment, I wrote down the same information that was on the service plan on the back of a brown lunch bag and asked if she would sign it. She was happy to, and when I got back to the office, the legal department was willing to accept it.

This is an example of the creativity and flexibility that are such important strands of synergy. Working synergistically doesn't mean throwing out the rulebook altogether; in the context of the United States, this is simply impossible. What it means is being able to see beyond the rules to the healing work that social workers and psychotherapists are meant to be doing. Instead of thinking like the system, with its checklists and forms and legal agreements that so often just serve to disempower both practitioners and clients, you need to think about the needs of the human being in front of you. What is best for their well-being? What can I do to ensure that my encounter with them will help unlock healing energy so that it can flow outward?

The other thing this story illustrates is the fact that those of us in the helping professions need to bear in mind that many of the people we will interact with have very good reasons not to trust us. It is hard for many clients not to see every social worker as a potential threat, a person who can take their children away or cut off their access to services and support. In my experience, too many social workers and psychotherapists expect clients to be grateful and obedient and become frustrated when the living, breathing people in front of them don't conform to the kinds of behavior that are most convenient for the system. Sometimes, they even feel personally offended: From their perspectives, they are nice people who are just trying to help.

This is why truth and reconciliation are such an important aspect of synergistic work in places like the United States. We need to understand that we are often not starting from a place of neutrality but of fear and distrust. Thinking back to my first meeting with Kitso, I am reminded of his ability to see my wish to learn about myself as well as about the profession, and the intentional way he created an environment to support my learning. If we translate this into the context of the United States, it becomes the responsibility of the professional or those in a leadership role to create an environment where trust can be fostered. It is our responsibility to reach out the hand and do the hard work of gaining the client's trust, understanding that they have good reasons not to trust us. If we're really going to do what we say we're going to do, which is to help, we need to be able to adapt to their needs and their perspectives rather than imposing our own.

As my career in the United States progressed, I had many opportunities to apply what I had learned from Kitso Mbenge—not by reproducing his exact methods, which are part of the cultural fabric of the world he lives in, but by seeing how the synergistic principles he operates by are applicable everywhere. The field of Black psychology undoubtedly helped. The scholars whose work laid the foundations for the specialist field of Black psychology, such as Drs Wade Nobles, Maulana Karenga, Na'im Akbar, Linda James Myer, and so many others, have added to my knowledge in the field of Black mental well-being, and reinforced that there is an alternative way to practice that takes into account the interconnectedness of the world we live in—despite the fact it has not been recognized by psychology more widely. It also created avenues for exploring how traditional wisdom can be brought into conversation with contemporary practice in the United States. For example, I had the privilege of attending the Association of Black Psychologists conference in Houston, Texas, in 2017, where permission had been granted by African healers to perform traditional rituals outside of the continent at the conference. Speaking to several early career professionals at the conference, many were confused, angry, and upset by the clear departure from the "established professional conferences" they had expected—but I and many others saw an effort to reclaim, restore, and promote the healing energy which Kitso shared so freely with me many years ago. We are still a long way from building a system that can even *comprehend* what synergy is, let alone practice it. And as long as we refuse to learn from the Indigenous healers who are the world's first psychologists, social workers, and medical providers, the question that needs to be asked is whether our current systems can actually give us the change we are hoping for: real progress toward well-being. If we keep doing the same thing, we will keep getting the same results.

Traditional healers like Kitso Mbenge exist all around the world, and they are rooted in Indigenous traditions of knowledge and practice. These traditions are, at their core, synergistic in that they value the interconnectedness of all living things and honor the fact that each of us has our role to play. What Kitso described to me is an approach to healing that respects the whole person and which demands much of both healers and clients. The healer must be present spiritually as well as physically in order for healing to happen, and the client needs to be prepared to participate in their own healing. There must be an orientation toward growth on both their parts.

But can this synergistic approach ever find articulation in a healthcare bureaucracy? We will explore the answer to that question in the next chapter.

Reflection

Peter Cornish

How might we bring more heart and soul into our modern healing practices? That is the question that comes to my mind when reading Gillian's account of her work in Botswana. In our efforts to disrupt and transform current mental health practices, we have applied implementation science and change management processes to encourage co-design and collaboration of all stakeholders. Typically missing from these processes is a method for grounding the learning within a spiritual context, a mission of service, or a larger sense of purpose and meaning. Major system building or transformation must incorporate healing practices. Large-scale change is called for to reconcile deep differences and conflicts. Mental healthcare practices have a long and continuing history of oppression tied to socioeconomic agendas aimed at maintaining power structures. As long as these structures are in place, the agenda will remain. While this is discouraging, committing to ongoing healing practices that incorporate spirituality, grief, and existential grounding in-community makes the lifelong journey of transformation and reconciliation sustainable.

Kitso's teaching for Gillian was about embracing a holistic, spiritually grounded, and community-oriented approach to healing, one that recognizes the importance of personal calling, humility, trust, and ethical responsibility. It challenged her to go beyond Western paradigms and adopt a more integrative, synergistic approach to therapy, fostering deeper connections with her clients and her own sense of purpose in the healing process. What lessons can I welcome from this chapter and its author, Gillian?

One possibility, as a descendant of European colonists, is to seek a deeper understanding of the peoples and practices that existed long before my family settled on the land where we spend summers—land that still holds significance for Indigenous communities. What lessons can I learn from those who came before me? At first, I didn't realize that this question would lead me on a journey back to something far older and more profound than I had imagined. It would not be about assuaging any sense of guilt but about genuinely connecting with a legacy that goes beyond my personal history. Gillian's own path of discovery prompted me to reflect on my own, reminding me that reconciliation is not about seeking absolution, but about engaging with something far deeper and more enduring.

For over four generations, my family has enjoyed the use of a rocky island in northern Ontario that holds a special place in my heart. The buildings have grown and expanded over the years, but the smooth stones of the ancient granite reefs that stretch out into the open bay and up and down the coast have been the same for time immemorial. It is a place that I know, at some level, connects me with something much larger than myself. Gillian has attended my family retreat several times, and the second time she came, in 2018, I realized that only one other Black person had ever visited the area. The cottagers were all white members of the upper or educated classes. It was at that time that I began to explore the privilege inherent in my relationship to the place.

When I confided this to Gillian, she regarded me with a bemused expression, her brow furrowed and eyebrows arched in disbelief. "What makes you say that?" she asked, her tone laced with incredulity. "You know, that whole White guilt nonsense is tiresome!" She went on to explain that while economic privilege is undeniably real, it is not the truest form of privilege. True privilege, she insisted, lies in wisdom.

I couldn't quite grasp it. Did I possess the kind of wisdom she was referring to? I wasn't sure. I wrestled with the unsettling shift I'd felt in recent years. Just a few days into my annual visits to the island, a once-vibrant connection to the land seemed to fade—the resonance with the trees, the rocks, and even the ripples in the water, slipping into silence. In its place, a quiet alienation would settle in, entwining with a profound desolation and a solemn ache, as though the land I had once known so intimately had quietly withdrawn, leaving me to mourn its absence in solitude.

Yet, for Gillian, the island was a sanctuary, perhaps even more so than it had ever been for me. She urged me to reflect on the significance of the place, on how it had shaped my values, my sense of connectedness, and my work in transforming mental health systems. Doubt lingered within me. But then, unexpectedly, everything shifted.

As I noted in my reflection in Chap. 5, in October 2023, I hosted a retreat for some of my Berkeley colleagues and my Stepped Care Solutions associates. All attendees had been involved in various aspects of mental health care system transformation over the past two years. At the time, I proposed that my Berkeley colleagues participate in a workshop on synergy, led by Dick Katz, my earliest mentor, to explore how it could strengthen our work. Dick suggested that since synergy had shaped both our lives in such intertwined ways, my own history needed to be part of the experience. As I prepared to host my colleagues in a place with such deep personal and family significance, I began to realize that I would be seeing it through the eyes of strangers, who would offer fresh perspectives. I couldn't have anticipated how much this would deepen my appreciation for the true value of the island.

Many things happened over the three days of our retreat. Plans changed and changed again due to travel delays and the challenges of transporting people to and from the island. The weather itself required us to readjust our schedule. It was an adventure, one that reminded us of the importance of openness and flexibility. Some of us got wet, and all of us learned more about synergy as we absorbed the stories Dick shared of the Indigenous healers he had spent time with. My colleagues were introduced to me in a new way. My family history was revealed in the land, the rustic cabins, the lapping waves, the intergenerational portraits, and the myriad of paths, trees, and rocks that form a special place in my heart. We spent three hours in a large talking circle, witnessing the experiences, stories, and spirit of everyone in attendance. While I was unsure of this at the time, we experienced the messiness of synergy without being able to capture it succinctly. I know this now because those in attendance have confirmed it through their continued conversations about the event and how important it will be to somehow continue the work we started.

Those three days in Ontario were profoundly spiritual for me, a quiet return to something I hadn't realized I was longing for. In preparing for the retreat, I reconnected with Kay, a local Indigenous woman who had once been very close to my mother. My mother had spent the last 15 winters of her life on the island, living without electricity or running water, immersing herself in the land's raw simplicity. It was with this same reverence that I wished to honor the island's legacy and its first stewards. I shared with Kay my intention to host the retreat there, hoping to tread lightly and respectfully in the footsteps of her people. Her face lit with warmth as she told me that her best friend, Marilyn, was an Elder of the Shawanaga First Nation, the original guardians of the land. Kay introduced me to Marilyn, who graciously agreed to offer a prayer to open the formal talking circle we had planned. Only weeks before, I had experienced a transformative moment, which I described in my reflection for Chap. 5, listening to an Elder at a conference—an experience that had stirred something deep within me. It felt as though Kay, Marilyn, and the land itself, with its ancient whispers and welcoming embrace, were drawing me back from the depths of my alienation and despondency, guiding me gently home.

For the first time, I saw how my work, my family, my ancestors, the land, and my colleagues were woven together in an ageless, interconnected web—one far greater than any of us could claim alone. This was spiritual synergy—alive in the people gathered, the ancient rocks, the enduring trees, and the endless expanse of the lake. It was home, a place where all things converged in quiet harmony. And in this moment, I realized that this, too, might be the wisdom Gillian was speaking of.

Reference

Bekerie, A. (1994). The four corners of a circle: Afrocentricity as a model of synthesis. *Journal of Black Studies, 25*(2), 131–149. http://www.jstor.org/stable/2784458

Chapter 7
Building Synergistic Systems of Mental Healthcare

Sarah Harte

Introduction

In the previous chapters, we have tried to make the case that synergy is not only an essential component of the healing process but also an ancient one that has been integrated into healing practices across many different cultures for millennia. If we want to develop less hierarchical, more person-centered approaches to healing, we need to find ways to make past wisdom modern by learning from traditional healers and applying their insights into synergy to contemporary systems of care.

At this point, we can imagine social work and mental health practitioners saying, "Synergy sounds good—but how does it work in practice?" It's one thing to see synergy at work in the filming of a documentary, as André Forget does in his chapter, or to gain an appreciation for the synergistic approach to training embodied by a traditional healer in Botswana. But what about the therapist working in a clinic in the downtown of a large North American city, who has to earn an income and answer to a licensing body? How can synergy be applied in the context of a clinic? In this chapter, I will attempt to explore just that.

A lot of thought went into the question of how best to do this. Up until this point, our book has sought to break with the standard tone of the academic monograph or research paper—previous authors have relied on storytelling, personal experience, and social theory to capture the nuances and complexities of synergy as an idea and practice. But while I have been inspired by their example to root this chapter in my own life and work, I also want to address readers who might be more comfortable with standard academic language, and so I have chosen to engage in a more hybrid writing practice, weaving personal and formal registers together. That means that

S. Harte (✉)
Stepped Care Solutions, Falls Church, VA, USA
e-mail: sarah.harte@steppedcaresolutions.com

P. Cornish, G. Berry (eds.), *Synergy*,
https://doi.org/10.1007/978-3-031-88493-1_7

this chapter will sound a little different from the previous ones, and will read more like a standard contribution to an academic book.

How we choose to write about something—the words we use, the tone we employ—cannot easily be separated from what we are trying to communicate. How something sounds shapes what it means. Our decision to shift registers here, so close to the end of the book, is intentional, and it is part of the larger conception of synergy at work in this text. While we value the power of the personal story, we also value the forms of knowledge that can flow out of surveys and quantitative data. Embracing synergy does not mean denigrating the tools and approaches typical of Western scholarship; rather, it means being open to the wisdom that is present in all traditions, including the Western academic tradition. It is therefore important that, in a book on synergy, we include a discussion of synergistic practice that works within the norms of mainstream social science. At the same time, I want to be open about the fact that I am approaching what follows not only as an observer, but as a participant.

I joined the leadership team at The George Washington University (GW) Mental Health Services in the spring of 2015. I was excited for the opportunity to step into a leadership role as the Assistant Director for Outreach, as I had previously worked in a handful of other university counseling centers in similar roles and recently relocated to Washington, DC. I had no expectations that this role and working experience would be transformational for me, as I was expecting the typical university counseling experience. I enjoyed working with young adults, and it was fun to be part of a campus culture, and the schedule and commute fit my needs. Little did I know that I would build professional and personal relationships that have lasted beyond the years of directly working together, and that my perspective on what it means to be a therapist, and to build community approaches to healing, would shift dramatically. I had not encountered the concept of synergy or Stepped Care 2.0 (SC2.0) prior to my work at GW, but as it turns out, working with these concepts and frameworks has been one of the most meaningful work experiences of my career.

In this chapter, I am working with data gathered on the implementation of SC2.0 at the GW Mental Health Services in Washington, D.C., to argue that it is possible to apply synergistic thinking and practice in the context of a modern mental health care bureaucracy. Stepped Care 2.0 is a model for delivering mental health care that requires the strands of synergy to function (the strands are: Truth and reconciliation, empathy, growth orientation, flexibility, interconnectedness, awareness, contextualization, creativity, diplomatic disruption, and agency). First developed in the province of Newfoundland and Labrador in Canada, it has since been implemented in several Canadian provinces and post-secondary institutions, as well as universities in the United States. Implementation at GW began in 2015, and this chapter highlights the role synergy played in implementation, and how this new system allowed for a more synergistic approach to mental healthcare practice in the context of a university counseling center.

This chapter is based on a thematic analysis (Braun & Clarke, 2012) of data from publicly available interviews with 15 key stakeholders held in June 2022, as well as my own experience as a clinician and part of the clinical leadership team working at

GW during this period. Of the 15 interviewees, three were university students, eight were clinicians or staff, three were clinical supervisors, and one was an individual in an administrative role. Interviewees were informed and consented to have their video-recorded interviews and transcripts used publicly by Stepped Care Solutions (SCS) for training and educational purposes. Interviewees shared their experiences over the course of two days, and interview questions were designed by SCS team members to shed light on each stage of implementation, as well as implementation outcomes. Later, in 2023, a member of the SCS research and evaluation team who was uninvolved in the original interviews with stakeholders conducted a thematic analysis (Braun & Clarke, 2012) of the interview transcripts to be able to highlight patterns in what stakeholders shared about their experiences.

Stepped Care 2.0 and the Synergy Solution

Stepped Care 2.0 is a model for mental healthcare that seeks to expand the range of treatment options available to clients, develop a system of care that is more responsive to client needs, and empower clients to take ownership of their mental well-being (Cornish, 2020). The model lays out nine steps of care from least to most intensive, with self-directed approaches to care in the lower steps, workshops and group programming in the middle steps, and specialist and acute care at the highest steps. When SC2.0 is implemented, these steps are populated through a co-design process to ensure that clients have a complete picture of all the mental health, e-mental health, and peer support services available in their communities. While the steps are laid out as a progression, they function like a buffet of options: A client who is receiving chronic care services (Step 8) could also be using interactional self-directed care through an app (Step 2) and intensive group programming (Step 6).

As a model, SC2.0 is rooted in one-at-a-time thinking (Hoyt, 2011), meaning that clients are able to choose the care that is best for them at the moment. Rather than signing up for a set number of treatment sessions and undergoing a lengthy intake process, SC2.0 treats each therapy session as if it could be the only session, emphasizing the importance of providing quick access to services while leaving the door open for clients to return as and when they need to. The one-at-a-time (OAAT) approach helps to ensure that rapid support is available for clients experiencing distress without overburdening the system. But it also reflects a philosophical difference between SC2.0 and other models for mental healthcare: Within SC2.0, the client is encouraged to see themselves as an active participant in their healing journey, and empowered to seek out care at the moment of need and readiness. This emphasis on client ownership over the healing process is enshrined in three of the underlying principles of SC2.0 (namely, self-direction, empowerment, and responsibility).

As a model for mental healthcare, SC2.0 upends a lot of conventional thinking about what the journey toward well-being looks like. We are used to thinking of clinicians as experts who apply their specialized knowledge by diagnosing a client's symptoms and developing a course of treatment that will lead to measurable improvement. In this paradigm, the clinician is the active agent who heals the passive client using their specialized knowledge: the doctor heals the patient. But within SC2.0, the clinician and the client are working in partnership, and each has their own obligations and responsibilities. This is a fundamentally different paradigm, one that relies on both parties being open to the workings of synergy.

The relationship between SC2.0 and synergy is complex. It is technically possible for an organization resistant to synergistic practice to follow all the steps for implementing SC2.0 on paper, and to end up with a system that can be described as following the SC2.0 model. But for SC2.0 to function as it is designed to function, it needs to fully embrace the strands of synergy. This is because the many interventions that make up SC2.0 are meant to work in concert in a way that produces a whole greater than the sum of its parts. Organizations that are drawn to SC2.0 because they believe it will help to reduce waitlists or direct clients toward lower-cost forms of care may be able to cut costs, but they will still be operating within the risk-averse, top-down, managerial paradigm of service delivery. Outcomes may improve marginally, but the fundamental problem that SC2.0 seeks to address—the overreliance on a professional class of healers with a vested interest in treating healing as a scarce resource—will continue. This problem can only be addressed through the adoption of a synergistic paradigm that acknowledges the role everyone—from professionals and clients to administrators and community members—has to play in promoting population-level well-being.

Synergy is the fuel that makes the model work, but the magic is not in the model. The magic happens when approaching the model and its implementation with a synergistic mindset that allows the model to thrive. At its best, SC2.0 opens channels for synergy to continue, to expand, and to come alive.

Implementation Background

The impetus for the implementation of SC2.0 at GW arose out of a tragedy: In the spring of 2014, there was a cluster of suicides on campus. Three students died by suicide, two of them doing so on the same day. This was a profound trauma for the community, and a system shock for the mental health services staff. Clinicians worked very long days to manage the demand for care, and the pain over the loss—and the fear that more students would die by suicide—forced the university to reconsider how it delivered services. It was around this time that the counseling center director heard about the SC2.0 model at a conference. Over the course of the following year, the counseling center prepared to transition to an SC2.0 model in the hopes that it would provide wider access and more rapid care to the most vulnerable students.

As discussed above, a key feature of the SC2.0 model is OAAT, which encourages students to seek help as they need it on a walk-in basis rather than going through a lengthy intake process and signing on for multiple treatment sessions. This is a more effective way of reducing wait times and guaranteeing access, helping to ensure that when students are experiencing a sudden crisis, they can get support in their moment of greatest need. OAAT is also consistent with synergy strands such as agency, flexibility, and creativity.

I was hired at the counseling center in the spring of 2015, and we started designing and populating our implementation of SC2.0 that summer. As the students left campus for the summer, the counseling team got to work on imagining what our SC2.0 services would look like when they returned. We created teams that were dedicated to re-envisioning how all of our services could work; everything, from how the students accessed services to the types of services that would be available. Some team members focused on clinician scheduling to allow for a walk-in open-access approach to care, while others began creating a group and workshop schedule. In my role as assistant director for outreach, I was heavily involved in writing content for our website and sorting out how to communicate this new approach to the student body.

This process was not easy. Not everyone on the team was sold on the idea that this model was the right fit for the center. Some staff were concerned about what it would be like to work within an OAAT mindset, worrying that they would never be able to do the type of therapeutic work they were trained to do and enjoyed, while others struggled with the focus on teamwork, having worked in the center for previous years where the work was fairly independent and disconnected. To enact all this change over the course of the three short summer months required a fast turnaround and a lot of pressure.

In the midst of our work that summer, there was a sudden and unexpected change in the leadership of the counseling center, and this crisis proved to be an opportunity for bold thinking. The change in leadership created a shift in energy that allowed the group to take a giant leap forward. Our new interim director, who had previously been one of three in a senior leadership team, now had the responsibility of guiding the team on her own. Fortunately, her clear focus on the work at hand and her leadership style of expecting the staff to take professional responsibility for our work, as well as keeping the focus on what is in the best interest of the students of GW, kept our development process on track. Because co-design is one of the core components of the SC2.0 model, one of the essential steps in implementation is establishing a sense of ownership among the staff who would be operating within the new system. According to staff who were interviewed, key elements of this process included sharing experiences and being transparent, leadership support, building relationships, finding a shared mission, and witnessing the benefits of the model.

Not all of the staff were supportive of the transition to SC2.0, however, and a number of mental health services employees left in November, only a few months into working under the SC2.0 model. A variety of factors may have played into these decisions, including a lack of faith in the leadership, skepticism about the new model, and reluctance to adapt to a new way of working. This meant that the staff who remained were, in general, the ones who were more open to the changes and more positive about the new direction.

Managing staff turnover is stressful. Those of us in the clinical leadership worried about overburdening the remaining team, and worried that we wouldn't be able to recruit new staff into such a "different" system. At the same time, I began to feel lighter. The staff who remained were more open to this different way of working. They were building trusting relationships with each other through authenticity and honest, flexible problem-solving. In the second half of the fall semester of our first implementation year, we started to observe some wins. We no longer had a waitlist, which was a relief for the staff who were responsible for managing it. Rather than worrying that the needs of waitlisted students weren't being met, they could focus on the students who were present: when clients came in needing help, we were able to say, "Come in, you can meet with a counselor today."

The turnover in staff also allowed us as a leadership team to get creative about how we could build a team that would thrive in this model. We began to identify skills and traits that could help a clinician work in a more synergistic way within SC2.0. We took an inventory of the special interests and training that our current team had, and made an inventory of the skills, interests, and training that was lacking so we could recruit candidates with those skills and interests. This was not limited to professional skills: We wanted to build a diverse team, a team that was made up of a variety of identities, races, cultures, interests, as well as different personality types and training backgrounds. Through this focus on hiring people for who they were as much as for what they knew, we began to build a team that truly became more than the sum of its parts.

Having secured buy-in from the remaining staff, it was necessary to engage with the student population who would be using mental health services. This was done by reaching out to the university community using marketing and communications, and making sure the students, faculty, and staff were informed about the programming and services available through this new model of care.

Community engagement and mental health education fell under my purview as assistant director for outreach. It was my responsibility to ensure that students, and the faculty and staff who supported them, understood how to access and utilize the services we offered. I approached this work in the typical way, by providing training to student development and residential staff, as well as faculty and other professional staff around the university. However, we knew that the typical one-way education or awareness campaigns would not change the culture alone. We needed to ensure that we were in direct communication with the students who we were there to serve. We worked as a group to engage students who typically did not seek out services—first-generation students, transfer students, and ethnically diverse students. We partnered with the multicultural center on campus to host events where we could receive feedback and engage in thoughtful dialogue about ways we could provide services that were truly supportive to more students. We met with the student government to hear their concerns about the services available, and asked for their feedback as we designed different services. We were changing our approach because we wanted the university community to feel heard, and we agreed that the old way wasn't working. When we shared this openly when we spoke with students, our authentic and honest approach created excitement and investment from the community.

Reflecting on the process of engaging the community, the administrator interviewed in 2022 stressed the importance of the relationship between buy-in and engagement, which, from a synergistic perspective, cannot be separated.

> I've been led by the philosophy of people supporting what they help create [...] I'm a firm believer in stakeholder engagement, and for us at a higher education institution, that includes a lot of different people. That includes students, they're our primary clientele, but it also includes parents, staff members [...] faculty, other administrators [and] partners across campus. And so I'm a firm believer in really gathering people together. [...] And so by engaging all those stakeholders, we have more ability to be successful because people felt they were bought into it. They were part of creating it.

Seeing leadership as a matter of partnering with individuals and groups rather than dictating to them is central to a synergistic worldview, but it can be difficult to integrate into a modern healthcare bureaucracy where authority is hierarchical. Building buy-in and a sense of ownership among staff is not, however, the same as giving every member of the team a vote or veto. Synergy does not mean leaders refuse to make hard decisions that could alienate some staff or community members. It means they practice transparency and accountability, seek input from others, and respect the interconnectedness of all communities. When systems are not working in a way that increases well-being, it is sometimes necessary to use diplomatic disruption to make them more effective. Managing the conflict that may ensue should be done with empathy and creativity, while leaders remain oriented toward growth on the individual, intrapersonal, and collective levels—including within themselves.

Challenges to Implementation

Implementing a new system is never easy, even when you strongly believe in the outcome. Along with the students, I had been frustrated by the status quo on all of the campuses I had worked at to date, and felt hemmed in by the arbitrary number of sessions (six to ten) that college counseling centers were offering at the time. I felt it was both too much and too little, and there often was not an effective middle ground. However, the scale of the cultural change we were confronted with was daunting. I had supervisees who were not happy about the new system coming to supervision each week, challenging me and the rest of the leadership about why we had made this change, and sharing their concerns with the model. I knew I would need to take responsibility for why and how we were changing, and find ways of addressing the concerns that were being brought to me.

Diplomatic disruption is one of the strands of synergy, and it is also an important part of the SC2.0 model. However, even the most diplomatic disruption is still disruptive, and when implementing a new system or a new way of working, leaders should anticipate challenges. In the case of GW, these challenges fell into five main categories:

Fig. 7.1 Implementation challenges experienced by supervisors, clinicians, and staff

- Skepticism
- Client expectations
- Fear
- Buy-in
- Paradigm shifts

Not all of these challenges had equal weight. We can see this in Fig. 7.1 (the Sankey chart below), which provides a visual depiction of how frequently interview responses coded under the various challenges cropped up, and the category of stakeholders who mentioned them. The size of the flows from the left to the right-hand side of the graphic represents how often respondents spoke about these issues. As you can see, interviewee concerns tended to cluster around two main areas: client expectations and the paradigm shift, with concerns about skepticism and buy-in being mentioned with less frequency, and client fear being mentioned only very occasionally (for this reason, we have opted not to explore it below).

Skepticism

Some supervisors, clinicians, and staff who reported feeling skeptical about the new model initially said they believed SC2.0 to be a thinly disguised attempt to cut costs. Others felt it was too different from what they had been trained to do. "My first reaction was like, what does this mean? What am I doing? I'm doing case management, I'm not doing therapy," said one clinician, describing their first reaction to the model. They had been taught that high-quality therapy meant regularly scheduled hour-long sessions, and the idea that they would see clients with less frequency seemed to fly in the face of this training.

One of the supervisors recalled being "very skeptical" of the SC2.0 model when they first heard about it, because they assumed it was designed to reduce the amount of care clients would receive.

> I had worked in college counseling centers before, and anyone who works in that sort of setting is familiar… with the reality that there are not enough resources for the amount of people who need your help. And so when I heard, you know, do less, give people less, all I could imagine was my client's responses to saying, actually, "You're saying you want all of these things, let me give you less than that" was how I interpreted it because of my training, and because of the past experiences that I had. [I] actually came from a center that had a ten-session limit. And we had a work culture where we were constantly trying to convince our director to give us just a couple more sessions, individual sessions with our students. And so, coming into a new setting where it was really encouraged to look at different steps and to consider what a client may need, I really wasn't hearing it. I was hearing that as "try to find a way to sell someone less than what they're asking for."

This response was understandable in the context of higher education, where cuts are often routine and where academics, clinicians, and other university staff have become accustomed to being asked to do more with less.

Client Expectations

One of the most significant challenges involved the established expectations clients had about what psychotherapy is. Like the clinicians themselves, many of these clients already had an idea of what it meant to go to therapy, and they expected treatment as usual. "A lot of the students that came into the center were really expecting what I had been doing in the past," one supervisor reported, "which was, I'm going to talk to a therapist and an individual therapy session for about 45 or 50 minutes an hour… and do that for a few weeks or a few months, and then I'll be all set and ready to go."

In some instances, expectation passed over into entitlement, and clients became angry that they weren't getting what they felt they deserved. A supervisor reported having to deal with

> students who were really entitled and felt as though they deserved or would automatically get what they wanted, in the form of an individual psychotherapy session once a week, for as long as they wanted it, by being a member of the university…. And sometimes, their parents would call too, and we'd have to tell them no, and help them understand why, and what they truly needed, and what the mission of the center was.

This challenge, however, provided a valuable opportunity for clinicians and supervisors to engage their clients in a dialogue about the purpose of therapy. When a growth orientation is embraced, diplomatic disruption can open the way for critical thinking and healthy questioning. One staff member said that after the initial shock of being told they would not get the kind of therapy they were expecting, clients were receptive to the idea that their concerns could be addressed more quickly. "That was usually received fairly well," the staff member reported, "because they just had never considered that."

Buy-In

Any larger organization contains multiple stakeholders, and getting these stakeholders on board with change can be an involved process. One stakeholder noted that a particular challenge lay in convincing the university leadership that the current system was not working. "[For] anybody who's in a position of power, accepting that their system is broken is a hard thing to learn," they said. "There were some blind spots there."

Another stakeholder noted that long-term success relied on being able to demonstrate in concrete terms that the new system was working by monitoring evidence and sharing progress. They said that "tracking wins and anecdotal stories and things that were successes along the way, along with the data—that's how we started to bring people along is that we started to see evidence throughout the years that things were changing."

Buy-in doesn't always happen overnight, and it's essential to have the patience to keep the conversation going and present evidence in a language that feels comfortable for key decision-makers.

Paradigm Shift

Most large companies and organizations do not work synergistically. Introducing a model of care that rests on the strands of synergy and requires a synergistic approach will feel revolutionary for those working within it, and revolutions are often uncomfortable things to live through. Because of its synergistic nature, SC2.0 isn't just another model for mental healthcare delivery that can be adapted seamlessly. It requires a change in thinking and practice on a fundamental level. It requires a paradigm shift.

Perhaps unsurprisingly, this was another significant challenge respondents identified. A supervisor reported tension between the Counseling Center, members of the psychiatric staff, and medical doctors who were not very well-versed in the SC2.0 philosophy and who "were not comfortable with… upending the status quo of the health and the healthcare delivery system."

Professionals who were trained to work in a certain way were sometimes reluctant to change how they worked, and afraid of what this new way of working would mean for them as therapists or social workers. "I think people get stuck in that traditional mental model of therapy," said one clinician. "Once they get past that fear, I think that it allows so much more and allows students to do so much more."

Acknowledging that synergy requires a paradigm shift is an important aspect of the process, because it allows leaders to anticipate the concerns stakeholders might have, and to avoid becoming defensive when changes are not immediately embraced.

Staff and Organizational Assets

The dominant paradigm in mental and primary health care tends to focus on deficits, challenges, and dysfunction rather than capacities, resilience, and assets. Synergy, on the other hand, respects and values the capacities and strengths we all bring, both to our own healing and the healing of others and to the systems, organizations, and communities we are part of. Implementing SC2.0 and encouraging clients and staff to adopt a more synergistic way of working involved challenges on a number of different levels, as we have seen, but in facing these challenges, they were able to draw on their strengths as well as manage the change. The success of the implementation was due in part to the assets clients, staff, and the organization itself brought to the process.

In interviews, the majority of respondents identified the importance of staff learning from others who had different expertise, knowledge, and experience than them as being key to implementation. Staff members expressed self-awareness of their own positionality and privilege and the importance of trying to understand the perspectives and experiences of others. "I think what was great is that we had such a large and very diverse clinical team," reported one of the clinical staff. "You could get feedback and hear from others in the group about how they were thinking about the model and what they were doing."

Cultural awareness, openness, and humility were also cited as being important assets that the team brought to implementation. Acknowledging that the counseling center was operating in a context where many clients had experienced forms of oppression that had the potential to shape their perception of psychotherapy meant that staff could see how a more synergistic approach to care could be beneficial. As one supervisor reported,

> What's important is our team was very aware of the… oppressive structures around us that would impact our staff and our clients and the work that we could do. And so, I think, being really critical and really strategic together from a disruptive standpoint, from an anti-racist standpoint, from a [standpoint of] making sure that we're doing what we know is right for our clients and our students, and as a team thinking about how we go about doing so when we're maybe in a system that would prefer we operate otherwise.

Alignment on core values like truth and reconciliation meant there was broad agreement about the fundamental project of building a more accessible and socially just mental health system.

Additionally, the director's leadership style fostered a "let's give it a try" attitude. Nothing was set in stone; we could always re-evaluate as the year went along. We revamped the schedule each semester and over the summer, asking ourselves what was working and what wasn't. If we didn't like an idea, we could give it a semester and see whether it succeeded or failed. This team flexibility is a parallel process to the clinical approach SC2.0 had introduced—to be creative and flexible, to trust your intuition, and to adjust as we went along.

Many care providers spoke of clients' and their peers' willingness to try something new as being a major asset for implementation. One clinician identified the "spirit of openness" as being essential to working within the SC2.0 model. "Everyone knew [that] at any point you were learning, and that you were growing, and that you actually needed to be continuously committed to that humility." This was true for clients as well as care providers, as another clinician pointed out:

> I think students are generally very receptive to change. You know, they're always on the bleeding edge or on the front line of [...] innovation themselves. And I've also been fortunate throughout my career to work with a lot of students who [are] willing to explore—you know, experiment or explore.

Together, these assets allowed the counseling center team to embody the growth orientation, contextualization, agency, and interconnectedness that are central to a synergistic way of working.

The individual assets of the team members were enhanced by the fact that they were operating in an organizational context that provided formal training on the SC2.0 model, as well as the opportunity for staff to spend time at a center that offered a version of stepped care so they could see what it looked like in action. "We all received multiple trainings on the stepped care model," said one clinician. "What it is, why it's important. So that helped me understand it." The care that was taken to ensure that staff had an opportunity to engage with and question the new model and to explore how SC2.0 would encourage their own autonomy helped staff gain a better sense of how SC2.0 worked from the inside and see how it would help solve problems in their workplace.

The Counseling Center also had other assets that proved a strength when it came to implementing a more synergistic model for care. Because it already had formal diversity, equity, and inclusion (DEI) structures and established roles directly tied to the mission of diversity to keep people accountable, conversations about the importance of social justice in mental healthcare were already part of the culture of the center. As one clinician put it,

> We had working groups that were based on diversity and inclusion. We also had working groups for training, and that actually highlighted making sure that we had multiculturalism and competent care across not just singular but also intersectional identities. And I think those are things that helped make sure that we, as a team, were being held accountable for really growing in diversity and our works that have, like, more of a DEI approach in general.

Because SC2.0 is animated by many of the same concerns that underwrite DEI initiatives in the workplace, staff had an opportunity to start reflecting on issues of privilege, social position, and systemic racism before implementation began.

Finally, implementation was handled in such a way that staff were not rushed into adopting the new model before they were ready. Approaching implementation with empathy, awareness, and flexibility meant ensuring staff had the time and space to grow into SC2.0 authentically. “We spent a lot of time just being ourselves, allowing for space, for people to say what their doubts are and what their concerns are, but also identifying some wins,” said one supervisor. “We spen[t] a lot of time saying that, like, it’s okay to question this, let’s watch the space, let’s see what happens.”

Organizational and System Outcomes

One of the biggest changes we made to the system was reworking our clinical schedules to include eight hours of “walk-in duty” or “triage” for each therapist, broken up into two four-hour shifts each week. This was a significant change. To access services, students simply needed to come into the center, and they would be seen right away by one of the therapists on walk-in duty, as opposed to the previous method of calling and making an appointment. This was the most stressful change we made, but also the most impactful one. As the 2015 fall semester began, we were all nervous about how this new open-access model would feel. What would it be like to sit with a student who we had no “intake paperwork” on and get right into helping them? Would we be overwhelmed by demand? What would happen if there was a crisis or complex clinical need? These anxieties were palpable, and they lingered throughout the year.

For the therapists, these eight hours of open-access availability were the most stressful part of the week, and no one wanted to be on the schedule for walk-ins on Friday afternoons. However, we quickly learned that these sessions were not all difficult. It felt great to have a student come in with a timely need and work it through with them right away. For example, one student was worried about a difficult conversation they needed to have with a parent while home on school break, and they came in the day before flying home for break. The student was able to work through their feelings, and even role-play the conversation, all the while knowing that they could come back after the break if they needed to. This would have been impossible in the previous model, because there would have been no availability for an intake the day before a break, and this wouldn’t necessarily qualify as a “crisis.” These here-and-now problems and helping conversations were satisfying to both the students and therapists alike. It was eye-opening to realize just how many students fit into this previously un-served space of needing help with something *now*, while not being in crisis.

We also built in identity-specific open-access hours where we publicized that students of marginalized groups would be prioritized, and therapists who had additional training and/or shared identities would be available. This proved to be effective for students who were looking to connect with gender- or racially diverse therapists and who often had additional barriers to seeking care. The therapists who staffed these open-access hours were also thrilled to contribute to the well-being of communities they were part of and cared deeply about.

The implementation of SC2.0 at GW began in the early summer of 2015 and was fully functional by the start of the 2015–2016 school year. As the new model came into effect, stakeholders identified four main positive outcomes:

- Efficient use of resources
- Increased access to care
- Reduced wait times
- Creation of safe spaces

As Fig. 7.2 shows, supervisors, clinicians, staff, and the administrator all mentioned the more efficient use of resources with the highest frequency when discussing key outcomes from SC2.0 implementation in their interviews, as compared to increased access to care, meeting demand, and creating safe spaces. A more detailed look at their transcripts suggests that this was due in large part to SC2.0's synergistic emphasis on flexibility and agency, which enabled clinicians to meet students' needs more quickly, accurately, and authentically.

One of the advantages of stepped care models, more generally, is that they provide clients with a range of different options to choose from. Expanding the range of options does not reduce the amount of care available, but it does challenge clients to think differently about care. Previously, the counseling center had offered students a set number of free therapy sessions, which may have encouraged some to believe they needed to use all of their free sessions in order to get their proverbial money's worth, or to ration their sessions out of concern that they would run out later in the year. With SC2.0, there was a change in mindset: Students could receive as much help as they needed, but the focus was on long-term resilience, recovery, and well-being.

One of the student stakeholders talked about how liberating it was not to have to worry about using up their sessions too quickly. Describing how the model had worked before SC2.0 was implemented, the student explained that "they had a restricted number of sessions. And so that made it hard to utilize it as much as

Fig. 7.2 Organizational and system outcomes reported by leadership, supervisors, clinicians, staff, and students

possible. I did try to stretch out as much as I could, so that kind of limited it." Having seen the SC2.0 model in action, this student was quickly convinced of its effectiveness and expressed a desire for this kind of approach to be made available nationwide. The fact that clients could seek out as much care as they needed, when they needed it, was a big part of the appeal.

Another major positive outcome cited by two students was the freedom to try out different members of the team to find the right fit. "When I did get to the point eventually where I could choose who I went to, I think it really changed everything," said one respondent. "It allows you to go to who you feel the most comfortable with and who you feel like understands you the best." This was especially important for students who wanted to engage with a therapist of the same race and gender as themselves. "I wanted to speak to a Black woman, and I wasn't willing to go any further unless that requirement was met from the very beginning immediately," a student recalled. "I was really surprised because usually, like, therapy or counseling centers, for the most part, it's like they have who they have on staff." The flexibility that was built into the new system allowed students to exercise their own agency in seeking out the kind of care they knew could help them. Rather than being passed through a production-line process in which each therapist was treated as being interchangeable, SC2.0 created the grounds for a more synergistic way of working: The strengths and values of each therapist were recognized, and students were empowered to be honest about their needs. For some, this completely changed how they viewed therapy. "Having multiple options," one student said, "was, I guess, kind of ground-breaking in my experience with counseling."

Some stakeholders stressed that the efficiencies created through SC2.0 had improved the quality of care all students were receiving. Because clients whose needs could be met outside of weekly intensive therapy sessions were connected to resources better suited to their needs, it freed up time for clients with more complex needs. "I would say it helped increase their access as well," a clinician noted. "Because again, we're diverting students to all these different interventions that probably would be better fit for them. But then that cleared the way for students who really needed, who really needed help and really needed access."

Given that the new system was implemented as a response to an ongoing struggle to reduce the risk of suicide by meeting student needs more quickly and creatively, some respondents focussed on the pragmatic, system-level advantages that came from working with the SC2.0 model. Many felt it was simply a more realistic way of dealing with the demands placed on a university counseling center that lacked the funds and staff to provide weekly one-on-one therapy to every student at the institution.

> We're not going to get more [resources]. We could spend all our time focused on, 'We need more, we don't have enough staff.' [...] Or we can, as a team there, [say] we've got this much, and we have this much that we need to accomplish; what's a creative way for us to be able to do that? And being excited about that really changes the dynamic and helps people to avoid burnout.

The flexibility of the system and the emphasis it placed on client agency and interconnectedness (both of which are strands of synergy) encouraged a mindset among service providers and service users that placed individual concerns in the context of community needs. "When you're operating a more flexible system and working with someone on what they need, not necessarily what it is that they may want, you're treating the entire system as your client," said one supervisor. "And that means there's a more equitable distribution of those resources."

Finally, stakeholders pointed to the elimination of the waitlist as one of the most concrete outcomes of the implementation. The fact that students could be seen on a walk-in basis as the need arose, rather than having to wait to receive care, was reported as being a major source of pride by some team members. "It was so effortless and seamless," one of the students recalled. "I literally made the decision to go speak with a therapist, and the next day, I was speaking with a therapist—it wasn't encumbered by paperwork."

Synergy in Practice

Synergy is an idea, but it is also a way of working. This means it can be difficult to capture in numerical data or statistics, which is why these stakeholder stories about their lived experiences at GW are so valuable. It is my contention that only through the stories and experiences of people working in synergistic ways can synergy's distinctive qualities—and its potential for transformative change—be understood. So, what does this look like in practice?

The GW case gives us some insight into how supervisors and staff experienced the transition and the role synergy played in implementing the new system. When asked what they saw as the key benefits of SC2.0, stakeholders indicated professional growth, flexible and stimulating work, and empowerment, autonomy, and agency as the main advantages they saw in the new system.

Clinical Judgment

One of the paradoxes of the established mental healthcare paradigm is that clinicians are viewed as being at once highly educated professionals, and also as potential liabilities who must adhere to established modes of practice and be subjected to strict oversight. In a clinical setting, it is not uncommon for therapists to dedicate a lot of their time to showing that they are following the rules of best practice. This means that there are often quite rigid constraints on clinical judgment. Synergy, on the other hand, can only operate in a system where practitioners feel able to respond to their clients on a less programmatic, more human level. One of the ways SC2.0 allowed for this was by giving therapists more space to exercise their creativity and clinical judgment.

As one stakeholder said,

> I learned to use my clinical judgment, and I learned to rely on my clinical judgment and seek support. And I also learned how to supervise much better as a result of the stepped care model, because there's a level of trust and encouragement and empowerment and strengths-based focus that I received in the model that we used.

Synergy is rooted in the idea that each individual node in a network has a vital and creative role to play and that it will play this role well when each other node supports, affirms, and challenges it to grow. When agency, interconnectedness, creativity, and empathy combine, healing flows through the system.

Another stakeholder noted that synergy provided a necessary corrective to the tendency that professionals sometimes have to grow complacent in their skills.

> I don't think there was a single day that went by where [I] didn't feel like I was uncomfortably, but necessarily challenged towards that growth and that humbling, that humility that is necessary, I think for our field, because we are holding people in their healing journey.

Synergy required more growth from practitioners, but it also provided more freedom.

Teamwork

In the dominant paradigm of the bureaucracy, each individual part is imagined as a cog in a machine that can easily be replaced with another, identical cog. Creativity, agency, and vision are reserved for directors and other executives, who are often lauded for their intellect or intuition and rewarded for their output, regardless of how successful they actually are in fulfilling the purposes of the organizations they lead. Unlike the assembly-line model that dominates most modern institutions, which seeks to produce uniform outcomes by controlling every input, SC2.0 seeks to affirm and develop the skills of each individual healer, creating a system of care that relies on the unique strengths each individual brings.

This means that the system is not owned by the leadership but by the clients, staff, clinicians, and supervisors, without whom it would not exist. Once embraced, this collective ownership empowers everyone to take their work more seriously.

> You really understand that, like you're able to do a lot more, and you're feeling empowered to really grow on your own skill set and say, oh, it really goes back to even owning the whole system. Right? Because now it's yours to do something with it as well. And to contribute to it. It's not just a system that someone else set up for you.

It also means that the system itself is not imagined as a set of rigid parameters that must be maintained but as an ever-responsive network where creative solutions can be explored in the moment, rather than having to be passed up the chain of command and signed off on by the person at the top (who will also, in many cases, take credit for it). Rather than a machine, the synergistic healthcare system is imagined as a body.

For instance, sometimes therapists would have a very busy walk-in shift and not be able to see all the people who arrived for services (this had always been one of our biggest anxieties), but turning students away was not an option. In times like these, teamwork was the only way to get through. We built a culture where there was an expectation that therapists would ask the team for help. Though there was initial hesitancy around this (surprisingly, therapists are not always great at asking for help), as we became more comfortable as a team and more trusting of each other, we found mutual accountability and support in teamwork. The experience of having a lot of students to manage in a shift and needing to ask for help made me more open to helping my colleagues when they, too, had an overwhelming shift. As our director would often say, "Flexibility goes both ways."

Flexibility

The SC2.0 model establishes a context where synergy can happen, but simply implementing the model does not magically create synergy—in fact, viewed dogmatically, the model can block synergy. A synergistic way of working recognizes that things are constantly evolving and is open to change in line with new needs and new developments. A synergistic implementation of SC2.0 recognizes that implementation is never really done.

For some stakeholders, this sense of constant evolution was seen as one of the best parts of the model as implemented at the counseling center.

> I love that it was always evolving. And [our director] gave us the leeway to interpret the model the way that we best work as well. Thinking about my own theoretical orientation, the way I show up in the world, you know, [just] use your strengths and where you come from as well to interact with your client populations.

Working this way requires a lot of trust. Clients need to trust clinicians, clinicians need to trust clients and supervisors, supervisors need to trust clinicians and directors, and directors need to trust everyone under them. But when these habits of trust have been developed, it is possible to make real progress toward measurable results like reducing waitlists, improving client outcomes, and increasing staff retention.

Once individuals and clients see how beneficial this approach is for their own practice and mental health, they are likely to become attached to the freedom and flexibility it allows. One clinician noted in their reflection that.

> the stepped care model, when that came in, that was quite a change because you had so much more flexibility and you had so much more control. And that was the thing that I enjoyed about it was having that flexibility, having that control, and for the client to have that control because a lot of the clients, you know, they were used to the old model as we were.

The greater flexibility of the model took some getting used to, but its results were immediately felt by the students we were there to serve. "I remember there being several options, and the option that I opted for was one-on-one," one student noted.

"I have done group therapy before, and I don't want that experience… I love the flexibility that was given to me, and the options that were presented. I had choices, and who doesn't like choices?"

Beyond Diversity, Equity, and Inclusion

The GW campus was and is a very diverse place. From the beginning, it was understood that diversity needed to be at the heart of what the counseling center offered. However, DEI concerns also needed to be understood in a synergistic way. There has sometimes been a tendency (especially in top-down, reactive approaches to DEI programming implemented by leadership teams that only have a shallow understanding of how different elements of our experiences as gendered and racialized individuals intersect with each other) for "diversity" actually to erase what is distinctive about each individual person. A synergistic approach to DEI sees the whole person and values everything they bring to the table.

One of the supervisors summed up the synergistic approach to diversity in the following way:

> [Diversity] was really important to us, not only around identity diversity, whether it be racial identity or gender identity, or other kinds of identities, but also about personality styles, different therapeutic styles. We really had a very clear message, which was, everybody just gets to be who they are and that there's not one cookie-cutter way of doing it.

In practice, this meant acknowledging the needs of students who had a strong preference for engaging with a therapist who shared some elements of their own demographic background (as seen above with the student for whom it was important to see a therapist who was a Black woman, and our model of specific walk-in hours for certain groups). It also meant being intentional about recognizing how identity was present in group therapy spaces. A student remembered how important it was that different needs were balanced in the group therapy sessions they attended: "They wanted to make sure that it was a safe space for everyone, even though they obviously want everyone to be open. They also didn't want your freedom of expression to then hinder someone else's comfortability, which was beautiful to experience."

In some cases, synergistically recognizing DEI principles in the therapeutic space meant helping students connect to forms of support that existed outside individual counseling sessions. One clinician recalled meeting a freshman student in a walk-in session who was struggling with some realizations about their sexual identity. "In that very limited time that she and I had together, we were able to really identify that she needed community support," the clinician said. "We had an LGBTQ group at the time, and I pitched it to her, and she got really excited about it." This is one of the principles essential both to synergy and the SC2.0 model: Healing doesn't just happen between the practitioner and the client. It is something that flows out of community and fellowship. Sometimes, the best way to help a client is to point them toward non-clinical resources that can help them explore their identity.

Truth and Authenticity

One of the most interesting interview quotes was from a student who talked about how helpful it was to be able to engage with clinicians who challenged them to think more deeply about their own behavior. "The individual counselors, the ones that I liked the most… did something that you typically don't see when you think about counseling," the student said. "They pushed me; they pushed me out of my comfort zone. They asked me questions that I didn't necessarily like or want to answer."

The student went on to say that their experience at the counseling center encouraged them to think about "not just what everyone else has done to you, but how your own actions have contributed to where you are in that moment." This can sometimes be one of the more difficult parts of the therapeutic relationship, especially in the context of one-at-a-time, where efficient use of time is of the essence. But for this individual, the experience of being pushed was profoundly empowering. Truth-telling and authenticity, on the part of both clinicians and clients, is an essential aspect of synergistic practice.

The importance of truth-telling and authenticity goes far beyond the clinician–client relationship though—in a sense, synergy begins and ends with our ability to hold nothing back in our encounters with others. When we do not feel safe to share our true feelings, critiques, and insights, it is very difficult to create conditions where the whole can be greater than the sum of its parts. Just as a therapist sometimes needs to speak hard truths to their client, and directors need to be firm with their team members, clients and team members need to have the courage to push back and use their own critical faculties. One of the dangers of hierarchical ways of thinking and practicing is that they impoverish rather than enrich the community: Arriving at truth is a collective process, and when only one person's voice matters, we end up with a distorted picture of reality. This ability to check ourselves and each other, and to engage in good-faith conversations in which we acknowledge that we can only see part of the picture, is a cornerstone of synergy.

Conclusion

A mental healthcare system that is meant to function synergistically can only be nurtured by planting seeds. It cannot be imposed from above, it cannot be fast-tracked, and it ultimately relies on the participation of each individual at every stage of the process. This can be frustrating for leaders who have a grand vision of how things could be, but have to work within existing structures and the personnel who have been trained to uphold them. And yet, there are no shortcuts: Synergistic outcomes rely on synergistic processes. An organization that needs to move toward a more synergistic model of care should develop a strategy for change that is itself synergistic in its approach to transformation.

This case study provides a glimpse into how the strands of synergy can transform healthcare bureaucracies. Even within larger structures that operate according to the logic of capitalism, risk, and scarcity, it is still possible to adopt more synergistic ways of working that can profoundly impact the quality of care provided to clients and the well-being of the people providing it.

Several lessons can be learned from the experience at GW. The first is that change is uncomfortable and needs to be approached with foresight and care. Preparing the ground for change and ensuring that staff and clients have the opportunity to learn as much as possible about what these changes will mean for them will help ease the transition. Opening up space for questioning, debate, and honest reflection gives stakeholders a chance to bring their own wisdom and experience to the process. At the same time, it is important that these conversations take place in the context of a clear vision from a leadership team dedicated to pursuing the plan. Debate should help the change evolve, but it shouldn't be a veto that keeps change from happening.

The second lesson that can be learned is that conflict should not be avoided. The implementation of SC2.0 coincided with a larger change of direction that alienated some members of staff. While it is regrettable that some staff chose to leave the counseling center because they felt the new model was too different from the way they had been trained to work, recognizing agency means acknowledging that not everyone will be ready to embrace change. Building a core team that understands the value of synergy and is willing to do the work required to enact system-level change is essential.

The third lesson is that synergy requires a paradigm shift, and paradigm shifts rarely happen quickly. When confronted with the unknown, human beings tend to fall back on familiar patterns of thinking and action. Mental health professionals understand that learning to relate to ourselves and others in new ways is hard work and that it takes time. Clients and practitioners alike need to see how the new paradigm works before they can trust it. Sometimes, changes in belief can only happen after changes in practice have shown us that a better way is possible.

In their interviews with SCS, staff and supervisors at the counseling center at GW were honest about their initial misgivings and spoke candidly about the challenges of growing into the new model. However, all said they felt that the process was worth it. Unfortunately, there is a coda to this story: In 2017, there were changes in senior management at the counseling center. New executives decided, without a thorough consultation, that the counseling center needed to return to a more business-as-usual approach. Under the guise of reigning in spending, changes were made that undermined the SC2.0 model. A hiring freeze was instituted, meaning that vacant positions could not be filled, and staff pay was frozen as well. Confronted with leadership that seemed overtly hostile to SC2.0 and unresponsive to staff concerns, many of the clinicians and supervisors who had built the new model chose to leave. It was not long before the counseling center reverted to a classical model of weekly one-on-one sessions, and waiting lists skyrocketed again. Synergy works.

But it is also vulnerable to one of the most entrenched elements of modern bureaucracy: authoritarian leadership.

This is not where the story ends, though. The clinicians and supervisors who left GW for new positions have taken the idea of synergy with them, and many are currently advocating for more synergistic approaches in their new workplaces. Bottom-up change takes longer than top-down enforcement, but it can also be more resilient in the long run. Given the existential crisis that modern mental healthcare faces in the form of rapidly expanding demand and limited resources, we are confident that the case study at GW is a forerunner of things to come.

Reflection

Peter Cornish

Sarah's story of the experience at George Washington makes me think of a metaphor—sowing the seeds of synergy. As was the case with Memorial University, where I first implemented SC2.0, the model was not sustained at GW. There was no fertile ground in either place at the time. Nevertheless, the seeds of synergy sprouted far and wide. The staff who nurtured synergy at GW have taken those seeds with them, transplanting them to other gardens. I have done the same in my work at Berkeley. Taking the metaphor further, the seeds are ancient, organic, and not genetically modified to produce controlled or predictable results. Synergy contributes to healing in unpredictable ways, but we can rest assured that it is a genetic powerhouse capable of growing and surviving in the harshest conditions.

This chapter outlines practical applications of synergy, but it comes with a warning. On the one hand, it illustrates what it looks like in clinical settings and in the learning required to make changes in the care system. It names five challenges one might anticipate when implementing SC2.0 (skepticism, client expectations, fear, buy-in, paradigm shifts) and suggests four possible outcomes if these challenges are met (efficient use of resources, increased access to care, reduced wait times, creation of safe spaces). The GW story offers lessons learned and questions unanswered. The chapter illustrates how synergy in practice guided the implementation of SC2.0 through the application of clinical judgment, teamwork, flexibility, a collective approach to DEI, and a process of truth, authenticity, and reconciliation. But the GW story is incomplete—a failure at the institutional level, while a life-altering success for the change-makers. The paradox resolves when we step back from the zero-sum binary of success/failure rooted in modern economies.

On the other hand, as Sarah suggests at the beginning of the chapter, something shifts (and, I believe, is lost) when we move to the "practical" and observable. The chapter serves as a cautionary reminder that in our efforts to bridge or braid the ancient spiritual practices with scientific, modern, and bureaucratic processes, we risk losing the foundation of well-being. Synergy involves recognizing the value of different analytic tools and languages without privileging one over the others. But this creates the risk that the deeper truths won't be understood—a tension that isn't always easy to navigate. It may involve awkward shifts in tone, such as the one we've seen moving into this chapter. The tension in the dance between old and new ways of writing and speaking is our reality and, likely, our destiny.

When I first read "synergy requires synergy" in this chapter, I did a double take. Was it a typo, a play on words? If the latter, it was too clever for me, and I may need Sarah to elucidate. Instead, I gave it some time. A seed germinated. Synergy is all around us, I remember learning from Dick. We first need to see where it already exists, and then by drawing on it just a little, we find it expands. Synergy requires tapping into existing synergy. It will survive and thrive as ecosystems adapt, perpetuating ever-evolving life. And if we attend to it respectfully and sustainably, our efforts will add to the expansion.

As with all things spiritual, synergy remains elusive, and the descriptions provided here are highly abstract. Any deep spiritual understanding is difficult. Tapping into synergy may, on a smaller scale, resemble the pursuit of Buddhist enlightenment or nirvana. Both are considered challenging to attain and grasp for several reasons:

1. They are complex. Enlightenment involves deep understanding of concepts like non-self (*anatta*), impermanence (*anicca*), and suffering (*dukkha*). These ideas can be counterintuitive and difficult to fully comprehend, especially when they challenge deeply held beliefs and perceptions.
2. Our minds are conditioned by habits, desires, and attachments that are deeply ingrained. We have a habit of wanting things to be observable and practical. Overcoming these habitual patterns requires significant mental effort and self-discipline.
3. Tapping into synergy, or attaining enlightenment, involves transcending the ego and attachments, which are central to our sense of identity and reality. Letting go of these can be profoundly difficult because they are integral to how we experience the world and understand ourselves.
4. Like the path to enlightenment, synergy is often described as a long and gradual process that involves extensive reflection, ethical living, and wisdom. The journey requires persistent practice and dedication over many years, or even lifetimes.
5. Like enlightenment, synergy entails realizing that the self is an illusion. Self is not an entity divisible from its context or the whole. For many, grasping this concept fully requires profound and sustained meditation or reflection experiences, which can be elusive.
6. Understanding and practicing the ancient teachings of Buddhism, or of Indigenous origin, might be difficult if one is not familiar with their cultural and linguistic contexts, which can make it harder to fully grasp the teachings.
7. Different cultures or societies interpret and approach the spirit or energy behind synergy and spiritual enlightenment in various ways. This diversity can sometimes lead to confusion or conflicting understandings about what synergy truly entails.

Thankfully, there are concrete ways to represent SC2.0. For example, the principles and the core components of SC2.0 have been codified, as described in our previous volume (*Stepped Care 2.0: The Power of Conundrums*). A framework for reimagining and organizing well-being resources was introduced in our first book (*Stepped Care 2.0: A Paradigm Shift in Mental Health*) and illustrated with many diagrams. The first book also reported on how the services changed and improved in a variety of settings. But neither book addressed how to sustain or expand the changes. The second book outlined the challenges. And this book opens a universe of possibilities without a map.

Healing on a global scale may seem unfathomable. I recall again the gallows humor of writer Dan Boyd in his climate crisis book, *I Want a Better Catastrophe*. Boyd is simultaneously realistic, fatalistic, and optimistic. At times, overwhelmed by whack-a-mole conundrums, my SC2.0 colleagues persevere, proclaiming with pride, "We are flying the plane while we build it!" John Norcross, in his foreword to the first book, said this transformation mission is not for the faint of heart. Nor is it for those without grace and humility. The phrase "on a wing and a prayer" originates from World War II. It was popularized by a 1943 song of the same name, written by Harold Adamson and James McHugh. The song was about a pilot's desperate situation, where the phrase "on a wing and a prayer" metaphorically described the pilot's situation as being precarious but hopeful. Mental health work is serious, and the myriad of crises facing us are overwhelming, but for that very reason, we can't take ourselves too seriously. We need luck, we need prayers, we need songs, we need humor. And we need people who are more practical than me and get things done! We need all of these things together. That is synergy.

References

Braun, V., & Clarke, V. (2012). Thematic analysis. American Psychological Association. In H. Cooper, P. M. Camic, D. L. Long, A. T. Panter, D. Rindskopf, & K. J. Sher (Eds.), *APA handbook of research methods in psychology, Vol. 2. Research designs: Quantitative, qualitative, neuropsychological, and biological* (pp. 57–71). American Psychological Association. https://doi.org/10.1037/13620-004

Cornish, P. (2020). *Stepped care 2.0: A paradigm shift in mental health*. Springer. Retrieved from https://link.springer.com/book/10.1007/978-3-030-48055-4

Hoyt, M. F. (2011). Foreword. In A. Slive & M. Bobele (Eds.), *When one hour is all you have: Effective therapy for walk-in clients* (pp. xix–xv). Zeig, Tucker, & Theisen.

Chapter 8
Building a Movement

Peter Cornish

Introduction

When academics study the history of ideas, they often try to discover the origins or roots of a concept in order to understand where it comes from. By identifying where an idea like democracy or spirituality originated, the thinking goes, we can learn something about its nature and value. At the beginning of this book, I told the story of how Dick Katz borrowed the term "synergy" from Buckminster Fuller, and then related it to the teachings he received from the Indigenous healers who mentored him around the world. I then explained how Katz introduced me to the idea when I was a graduate student at the University of Saskatchewan, and how my colleague Gillian Berry and I came to see synergy as the necessary precondition for Stepped Care 2.0 (SC2.0).

But synergy is not simply an idea—it is a practice, one that is essential to our social life as humans, and one that Indigenous peoples the world over have placed at the heart of their understandings of the cosmos. Synergy is always around us, latent in every human interaction, and from time immemorial, Indigenous peoples have understood the power of synergy to heal individuals and communities. The idea of synergy does not have a single lineage. It lives in the myriad stories people have told throughout the millennia to explain how it is that our existence on this planet is simultaneously individual and collective: A matter of profound uniqueness (the uniqueness of each person, culture, and species), but also and everywhere a matter of blending, sharing, and collaborating with each other and the natural world.

Doing justice to this powerful and ancient idea meant opening ourselves up to synergy itself, which is why the process that led to the creation of this final volume in our series about SC2.0 was a deeply synergistic one from start to finish. Our

P. Cornish (✉)
Student Mental Health, University of California, Berkeley, Berkeley, CA, USA
e-mail: peter.cornish@steppedcaresolutions.com

P. Cornish, G. Berry (eds.), *Synergy*,
https://doi.org/10.1007/978-3-031-88493-1_8

ambition was to return synergy to its rightful place at the heart of mental healthcare, and to do that, we felt it was necessary to move beyond the purely analytic and materialist modes of inquiry that are part of Western academic practice. By embracing a multivocal, storytelling approach, we sought as much as possible to write the book in a way that would honor the material.

This is also why we have tried to stress throughout this book that synergy is not a recently invented concept, and nor is it one that I, or the team at Stepped Care Solutions (SCS), can claim any ownership over. Academia, like the corporate world it is increasingly modeled on, rewards those who develop new and innovative concepts, models, or processes. One of the greatest achievements any academic can aspire to is to have a theory, idea, or discovery named after them. But there is something troubling about this desire to turn an idea into a "brand"—however necessary it might be to widen an idea's reach and unlock the funding that is vital to continuing the work. This is especially true when it comes to an idea like synergy, which is rooted in Indigenous ways of knowing, and which is powerful precisely because it recognizes the importance of community, collaboration, and collective action. Synergy is unique to each situation, and cannot be packaged neatly into a staged formula, or an acronym easily digestible to the mental health system as a whole.

We were inspired by the fact that Katz always dedicates time in his books and lectures to talking about his relationships with the Indigenous healers who taught him about synergy. Given the colonial history that has shaped so many of our assumptions about academic research, it is essential that we not perpetuate imperialistic traditions of cultural appropriation. We are not claiming, in this book, to have discovered synergy; synergy discovered us. We are simply trying to follow in the footsteps of Katz and his teachers = Oma Djo, Ratu Civo, Joe Eagle Elk, and Danny Musqua by bringing this concept to a wider audience, and other Indigenous teachers and guides who have shared their knowledge with us.

At the same time, we have been wrestling with how this idea of synergy might be operationalized in the context of the modern mental healthcare system—a system that has its roots in colonial ways of thinking and which, in its present incarnation, is highly institutional, materialistic in its philosophical assumptions, and risk-averse. In the first volume in this series, *A Paradigm Shift in Mental Health*, I argued that if we wanted to really meet the needs of the many people seeking to improve their mental health, we needed a new system, which I called SC2.0 (Cornish, 2020). That first book was structured like a typical academic monograph, and I was listed as the sole author. It was a deeply personal rallying cry against the entrenched problems I had observed over the course of my career as a psychologist working inside the academy, and it garnered a lot of interest. But while SC2.0 promised to open up the bottlenecks in mental healthcare by empowering communities to be more active in their own healing, I quickly discovered that some of the institutions seeking to implement SC2.0 did not really understand its most radical elements. This led to the publication of a second volume, *The Power of Conundrums*, edited by Gillian Berry and I, in which dozens of mental health researchers and practitioners explored how the model was being misunderstood (Cornish & Berry, 2023). The final chapter of that book included a roundtable in which three of us, Berry, myself, and our SCS colleague AnnMarie Churchill, concluded that something was still missing.

That missing element was synergy—which is why we dedicated this final volume in the series to the questions of what synergy is, and an attempt to explain why synergy is essential for the proper functioning of SC2.0. This was not an easy task. Synergy is something almost everyone has experienced at one point or another, but often in circumstances that are viewed as unique, like a community coming together in the wake of a natural disaster, or a team learning to trust each other during a corporate retreat. We wanted to explore the philosophical roots of synergy, seeing this as a key factor in the outcome, but we also wanted to show that synergy is not just an idea but an experience and process that can be fully incorporated into how we operate in the world of mental healthcare. Like all worthy expeditions, the exploration began with questions. How do we make synergy the norm rather than the exception? How might we bring synergy and synergistic thinking into the clinic, the community health center, and the office? And so we set ourselves the goal of writing a book that could simultaneously explain what synergy is while also showing how synergy and SC2.0 have worked in practice.

Writing About Synergy, Synergistically

Even at the outset of this project, I knew such a book could not be written by just one person. The essence of synergy is the coming together of different elements to create something greater than the sum of its parts, and just as my own understanding of synergy has been enriched over the years by my relationships and conversations with people from many different walks of life, writing a book using a dialogic, relational process felt like the only appropriate way to move forward. But this meant being open to risk, uncertainty, and sometimes a little bit of frustration. Should we try to couch synergy in terms that would be more palatable to an academic audience, or should we be willing to talk honestly about subjects like spirituality, which might turn some readers off? Should we dial back criticism of the existing mental healthcare system and the Western, Eurocentric cultural paradigm in which it operates to avoid ruffling feathers? How, as a group of authors with a variety of racial, ethnic, gender, and sexual identities, should we address the difficult questions of privilege and accountability that are inevitably raised over the course of this kind of collaboration? It was not always an easy process. We discovered firsthand that engaging with each other synergistically does not mean avoiding conflict, and that a synergistic process does not always arrive at clear consensus. At the same time, we learned that with enough trust, patience, and generosity, synergy can enable marvelous things.

We also realized that in addition to trusting each other, we needed to extend that trust to you—the reader. While we have tried at every stage to develop a text that clearly communicates what synergy is, where it comes from, and how it relates to SC2.0, we have had to embrace the risk that comes with authenticity. We know that not every reader will read the book cover-to-cover, as it was intended to be read, and not every reader will see equal value in all of the chapters. Every reader will come to these chapters with their own experiences, ideas, doubts, fears, and certainties. Just as this book was the product of a community of thinkers, we hope that readers will see themselves as participants in this community, too, and will extend our thinking through their own insights, stories, and critiques. If something in this book speaks to you, we hope you will enter into a dialogue with us as we continue to pursue the mission of well-being everywhere. Synergy gets stronger with every input.

Looking Back on the Journey

This is not a book that has been organized according to the traditional logic of the academic monograph. Rather than laying out an argument at the beginning and then providing a series of points that back up or expand on that argument, we have imagined this text as a journey. The first chapter introduces the idea of synergy as it appears in Katz's work, explains the relationship between synergy and SC2.0, provides an account of how evidence for synergy is imagined, and then gives an outline of what is to come. This is our first step onto the road that will take us deep into the philosophical underpinnings of synergy to its practical applications.

The second chapter serves as a backward look over the shoulder at the previous two volumes in the series, *A Paradigm Shift in Mental Health* and *The Power of Conundrums*. We had ended *The Power of Conundrums* on something of a cliff-hanger, having explored the deep challenges facing the SC2.0 model in theory and in practice, but having chosen not to rush into prescriptive solutions—instead opting to let the conundrums percolate and inspire creativity. In the second chapter of this volume, I return to these conundrums, arguing that they are each, in their own way, produced by a lack of synergistic thinking and practice. By revisiting the conundrums through the lens of the strands of synergy, I make the case for why SC2.0 requires synergy to function.

While the strands of synergy are the fuel that powers the SC2.0 model, and the SC2.0 model is designed to operate within modern mental healthcare bureaucracies, that does not mean that incorporating synergy is easy. The third chapter in the volume picks up the question of why synergy is so hard to achieve in contemporary institutions. Why is it that the conundrums we have identified are so hard to resolve, even for people who understand the power of synergy? Through wide-ranging interviews with three mental health professionals at varying stages in their careers, André Forget provides an account of why synergy is so radical in the institutional context, and how synergistic thinking can be successfully incorporated. By looking at the experiences of Dr. Genée Jackson, Dr. Raghav Suri, and Kareem South, this chapter shows the many different paths these thinkers have taken to arrive at the conclusion that mental healthcare needs synergy. Reflecting on political action, epistemology, and spirituality, these three researchers explore why Western institutional thinking has so often been hostile to synergy, while also sharing stories of how synergy has been made to work in institutional contexts.

Having introduced the basic ideas underlying synergy, connected them back to the conundrums of the previous book, and explored the institutional context that makes synergy hard to implement, we turn in Chap. 4 to a question that has particular contemporary relevance: Institutional attempts to overcome histories of violence and exclusion through mandates for diversity, equality, and inclusion (DEI). In "The Gaslighting of a Profession," Catie Greene weaves together academic critique with personal experience to argue that the mental health industry, despite its stated desire to redress the historic and ongoing violence of colonialism, has fallen short of the mark. Greene argues that while the intentions behind DEI initiatives may be well-meaning, in practice, they have often failed to achieve decolonization because they still operate within a Western bureaucratic paradigm. Synergy, for Greene, is a necessary corrective, one that, when authentically implemented, points the way toward truth and reconciliation.

But what does synergy look like in practice? Chap. 5 presents a travelog in which André Forget describes the filming of a documentary about the life and work of Richard Katz. Forget writes about Katz's life and ideas, and also provides a portrait of how a diverse group of academics, psychotherapists, and film-makers came together in the autumn of 2023 to explore synergy in a creative and collaborative way. By providing a behind-the-scenes look at the conversations and friendships that developed over the course of this trip, Forget seeks to understand what makes synergy so powerful, but also so vulnerable. Observing that the greatest danger to synergy lies in the fear that opening ourselves up to others will be dangerous, Forget concludes that synergy is predicated on deep honesty and trust that needs to be nurtured over time.

Chapter 6, "The Education of a Healer," is in many ways the climax of the book. Having been raised to believe that the Western way of healing is the only one available, Gillian Berry's experience studying with Kitso Mbenge, a traditional healer in Botswana, is truly eye-opening—but it also raises questions about how this authentic, spiritual form of healing can be applied in the very different context of the US social work system. Berry's willingness to wrestle deeply with the destructive legacy of colonialism, and the challenges of trying to build a more synergistic future for mental healthcare, gets to the heart of why decolonization is so important, and provides a rousing call to action for those who have seen how badly the current system is failing, and are serious about finding an alternative. Berry argues that Indigenous healers, our First Psychologists (Katz, 2017), should be our guides in charting the way forward.

The end of our journey comes in Chap. 7, with a return to the world of Western mental healthcare. Having explored what synergy means, seen examples of synergy in action, and gone all the way back to the synergistic Indigenous healing practices that colonialism has tried to erase, we are ready to confront the challenge of how to apply synergistic ways of working in the context of a modern clinic. Sarah Harte's case study on the implementation of SC2.0 at The George Washington University (GW) in Washington, D.C., highlights the ways in which this implementation was made possible by synergy, and, in turn, made a more synergistic approach to healing possible for GW students. Working with data collected in 2022, Harte's chapter explores the challenges and successes of the SC2.0 implementation, providing a compelling argument for how synergy can be unleashed even in the setting of a modern clinic.

We are grateful to the readers who embarked on this journey with us, and we hope that it has inspired them to think about the times when synergy has appeared in their own lives. Synergy is always present as a possibility, and when I think about my own life, some of the most powerful moments of healing I have experienced have, in one way or another, been the product of synergy. It is irrepressible. The question we posed at the beginning of this book is whether or not that synergy can be harnessed in a more systematic way so that its revolutionary healing energy can spread through our entire society. I believe we have answered it with a "yes!"

Building a Synergistic Movement

This book is the last in our trilogy about SC2.0. It signals not an end, but a beginning. The strands of synergy that we have developed at SCS are a guide to how we can move forward in the future. Perhaps more strands will be added in the future as we deepen our knowledge of synergy. We must take responsibility on the individual, interpersonal, and collective levels for the shortcomings of the existing mental healthcare system, which all too often fails to provide adequate care for the most vulnerable. In doing so, we should turn to the strands to help us cultivate the habits and practices that will enable us to make our accountability an agent not of self-recrimination, but of real and lasting change.

Truth and reconciliation is our north star, the ultimate goal of our work. If we are not moving toward truth and reconciliation, we are moving in the wrong direction. To practice truth and reconciliation, we need **empathy** for the experiences of others. We need a **growth orientation** to enable us to remain humble, open to hearing what others say, even when we disagree. We need to practice **flexibility** so we do not become rigid in our thinking, or make the mistake of becoming too dogmatic about our models and practices. We need to value **interconnectedness**, recognizing that the actions of each individual and their ancestors have an ongoing impact, for good or ill, on the people around them. No one is in this alone. **Awareness** of self and others helps us understand the complex ways in which privilege manifests and shapes our experience of the world. **Contextualization** reminds us that things often happen for complex reasons; even the simplest phenomena cannot be understood if the context in which they emerge is ignored. When faced with problems that seem intractable or insurmountable, **creativity** is essential. If a policy, process, or procedure is not working as intended, it might be time for out-of-the-box thinking. Conflict is an inevitable part of human life, especially when working as an agent of change. **Diplomatic disruption** provides a path toward re-imagining conflict as an opportunity for growth and an avenue for transformation. Finally, we must recognize not only our own **agency** as mental health professionals, but also the agency of every single client we engage with. Healing is a collective effort; when we engage whole communities in it, healing energy can flow from person to person until it changes the world.

I recognize that to some readers, this might sound a little bit too idealistic. Can synergy really transform the deeply embedded systems that underlie the operations of the modern world? This is a fair question, but I think it needs to be re-framed. Given the fact that we are in the midst of an unfolding climate catastrophe, where the needs of our population grow greater by the year even as the resources needed to meet those needs grow scarcer, operating within systems that are underfunded and stretched to breaking point, the better question might be "how can we possibly continue if we *do not* draw on synergy?" Therefore, it is our contention that only through the untapped energy of synergy—the energy of billions of humans who are not currently empowered to heal each other and the planet—can we possibly begin to imagine a better future. We invite you to join us in building a synergistic movement that will allow us to move toward a world in which healing energy is finally allowed to flow.

References

Cornish, P. (2020). *Stepped care 2.0: A paradigm shift in mental health.* Springer. Retrieved from https://link.springer.com/book/10.1007/978-3-030-48055-4

Cornish, P., & Berry, G. (Eds.). (2023). *Stepped care 2.0: The power of conundrums.* Springer. Retrieved from https://link.springer.com/book/10.1007/978-3-031-45206-2

Katz, R. (2017). *Indigenous healing psychology: Honoring the wisdom of the first peoples.* Healing Arts Press.

Afterword

Richard Katz

It's 1968. I'm living with the Ju/'hoan people of the Kalahari Desert, seeking to learn about their approach to community healing. Still living primarily as hunter-gatherers, the Ju/'hoansi offer a glimpse into the way we as humans have lived for nearly ninety-nine percent of our existence. I believe their approach to healing can offer us important, fundamentally sound, evolutionarily tested teachings. And what I experience in the Kalahari—which has since been confirmed by other experiences with Indigenous peoples—is an approach to healing that, to my Western-trained eyes, is both revolutionary and quite amazing.

In their healing dance (their primary community gathering), the Ju/'hoansi release a healing energy that encourages a community experience where people come together to create a greater whole from disparate, even conflicting parts. Respecting differences, a reality grows where "what is good for one becomes good for all," as the "whole becomes greater than the sum of the parts." Healing, as a valued community resource, expands, becoming accessible and renewable. The more healing that occurs, the more healing becomes available.

As someone trained in the Western approach to medicine, where healing is institutionalized and treated as a scarce resource, and where healing for one person means there will be less available for another, what I experienced in the Ju/'hoan approach seemed almost impossible. Instead of a competition for resources, with access favoring those already favored by their circumstances and privilege, I saw healing flow from person to person.

I realized that the concept and experience of "synergy" was helpful to understand this Ju/'hoan approach to healing—especially the idea that synergy already exists in the universe, and that it is our responsibility to allow it to exist in our human activities.

Thus was born my commitment to fulfill the request the Ju/'hoan asked of me: "Tell our story to your people." For the next nearly 60 years, drawing on my training and position as a clinical psychology Ph.D. with teaching experiences at Harvard and First Nations University of Canada, and the further healing work I've done in

P. Cornish, G. Berry (eds.), *Synergy*,
https://doi.org/10.1007/978-3-031-88493-1

places like the Fiji Islands, I've tried to channel synergy to impact the emergence and delivery of community healing in Western contexts. My main focus has been on those whose responsibility it is to deliver that healing—those who are among "my people."

It has not been an easy path. It's often a struggle to talk and write about synergy as a concrete and accessible source of healing. But there has been some wonderfully exciting and meaningful work with doctoral students, focusing on issues related to synergy and healing. Peter Cornish was one of those students: In 1997, he completed a stimulating and insightful doctoral thesis on synergy and social change. And just recently, I've completed a culminating point in those efforts, describing possible applications of synergy to Western contexts in my book, *Indigenous Healing Psychology: Honoring the Wisdom of the First Peoples.*

Around the same time I finished the book, I received an unexpected but welcome message. It was Peter Cornish—again impacting my life, this time through his Stepped Care Solutions organization, through which he is working with passion, creativity, and sensitivity to bring synergy into Western institutional settings, such as university counseling centers. "Are you interested in working with us?" he asked. Was I interested? There was no hesitation on my part. It was a chance to progress in my work into areas where on-the-ground efforts are having a definite effect. "Yes," I responded—just a clear and simple "yes." And so a new phase of my commitment to synergy in healing began.

And now I have in front of me this special book—*Synergy: The Healing Energy Guiding Stepped Care 2.0.*—and I'm filled with gratitude. I see how well the book describes ground-breaking work that has moved my own thinking about synergy and healing more into real-world situations, thus deepening its meaning. It gives me a feeling of passing along something that was given to me in order to give to others.

I'm honored to be asked to write this Afterword. I'm so impressed to come in contact with this book, to savor its depth and insight. Within the chapters' engaging discussions of Stepped Care Solution's work, I find a group of creative and dedicated scholars and practitioners who are doing the actual work of introducing synergistic principles into community healing settings. They are creating the reality out of the apparent paradox that "the more healing that is given, the more healing becomes available." They are offering a real alternative to the prevailing Western scarcity model of healing, where people must compete for access to limited healing resources, with those already having privilege and power having an unfair head start.

In the various and varied chapters, there is a careful and respectful focus on the elusiveness of synergy, going beyond efforts to seek only conceptual clarity in order to face the more fundamental challenge of focusing on the *experience* of synergy. We learn about the need to embrace uncertainty and the unexpected in order to allow the actual experience of synergy in healing to emerge.

Throughout the chapters, we hear the commitment to asking difficult questions about implementing synergy, questions about how to introduce a very elusive, experientially based concept into Western helping settings that typically overvalue conceptual clarity and so-called "objective data points" over lived experience. The gift of these chapters is both their refusal to offer easy answers to those questions, and their commitment to making this concept of synergy an actual, practical aspect of peoples' lives. Continuing to wrestle with these essential questions, they help keep them alive—and thereby invite us to open ourselves to what is a necessary element to change: a willingness to "not know."

This book fulfills an essential need to bring healing to all those who require it. Describing both ongoing work and commitments to future directions, the book shows how an essential approach to healing—which at first glance might be considered "unrealistic" or "unattainable"—can exist, grow, and evolve. Courageously, the authors have made a commitment to engage with the unknown, and to nourish an openness to and valuation of the unexpected. With their willingness to take risks for the greater good, they are living out the necessary conditions for this powerful experience of synergy to arise and thrive.

I am deeply appreciative of this book. Stepped Care Solutions has done and is doing the necessary work on the ground in healing settings for synergy to become a felt and practical reality—with the potential for healing to touch us all. This book inspires all of us to act toward this goal. And what more could we hope for?

Index

P. Cornish, G. Berry (eds.), *Synergy*,
https://doi.org/10.1007/978-3-031-88493-1

MIX
Papier aus verantwortungsvollen Quellen
Paper from responsible sources
FSC® C105338

If you have any concerns about our products,
you can contact us on
ProductSafety@springernature.com

In case Publisher is established outside the EU,
the EU authorized representative is:
Springer Nature Customer Service Center GmbH
Europaplatz 3, 69115 Heidelberg, Germany

Printed by Libri Plureos GmbH
in Hamburg, Germany